Neolocalism and Tourism:
Understanding a Global Movement

Editors:

Linda J. Ingram, Susan L. Slocum and Christina T. Cavaliere

(G) Goodfellow Publishers Ltd

(G) Published by Goodfellow Publishers Limited,
26 Home Close, Wolvercote, Oxford OX2 8PS
http://www.goodfellowpublishers.com

British Library Cataloguing in Publication Data: a catalogue record for this title is available from the British Library.

Library of Congress Catalog Card Number: on file.

ISBN: 978-1-911635-60-4

Design and typesetting by P.K. McBride, www.macbride.org.uk

Cover design by Cylinder, www.cylindermedia.com

Contents

List of figures

List of images

List of tables

Dedication

This book is dedicated to all frontline workers and their tireless efforts to provide safety during the COVID-19 pandemic as we endeavor to find a unified path forward.

We also dedicate this volume to the community members, artists, craftspeople, innovators, entrepreneurs, social activists, just decision-makers, small business owners and educators who are driven by the collective mission to conserve and reinvigorate histories, stories, and crafts that result in unique place-making around the world. May our home(s) continue to thrive in biocultural diversity!

Author biographies

Thiago Allis, PhD, is an assistant professor at the University of São Paulo, Brazil. His research interests include urban tourism, regional and urban planning, and tourism mobilities. Thiago has led and participated in several research and outreach projects funded by Brazilian and international agencies. In 2012, he was an invited lecturer in East Timor.

Jessica Aquino, PhD, is an assistant professor at Hólar University in the Department of Rural Tourism, and Head of Tourism Research at the Icelandic Seal Center. Her research interests are in sustainable community development and quality of life; sustainable tourism and responsible tourism practices; and responsible management of natural areas.

Miranda Bradshaw graduated with a BA in Economics from Boise State University and a MS in Applied Economics from Utah State University. She currently lives and works in Copenhagen, Denmark as a user experience (UX) designer and product manager for a medical technology company.

Christina T. Cavaliere, PhD, is a conservation social scientist. Her research involves the human dimensions of socio-ecological systems including tourism impacts and biocultural conservation. Dr. Cavaliere serves as an assistant professor at Colorado State University and has experience working with universities, communities, businesses, non-governmental organizations, and multi-lateral institutions on six continents.

Vorakarn Chalermchaikit is a PhD candidate at Assumption University, Thailand, majoring in tourism management, and a part-time lecturer at Dhonburi Radjabhat University. Her research focuses on gender issues in hospitality impacting both the hospitality and tourism sectors. Ms. Chalermchaikit has strong backgrounds and experiences in both domestic and international hospitality.

Yixuan Cheng, MA, is a PhD candidate, majoring in Tourism and Hospitality Management at Assumption University, Thailand. Ms. Cheng finished her Master's degree in International Communication and Culture at University of Leicester, UK. Her research interests involve sustainable community tourism.

Kynda Curtis is a professor in the Department of Applied Economics at Utah State University. She received her PhD in Economics from Washington State University in 2003. Her research interests include agriculture/food marketing, direct-to-consumer marketing, as well as consumer demand for value-added, specialty, and differentiated foods.

Ros Derrett, OAM, PhD, Adjunct Professor in the School of Law and Justice, Southern Cross University is an experienced senior academic/consultant who works with individual artists, cultural enterprises, businesses, communities, and local government in regional Australia to address ideas and issues connected with designing and delivering quality experiences for cultural tourism afficionados which provides community participation opportunities for dynamic business development and place-making.

Michael Fagence, PhD, is a geographer who has linked his training in 'thinking geographically' with elements of tool-kits from history and from semiotics to forge investigative and interpretive processes that have the versality to be applied to cases of history-linked and heritage-based tourism. His most recent appointment was as Honorary Research Fellow at the University of Queensland.

Linda J. Ingram, PhD, an adjunct professor with George Mason University, teaches sustainable tourism development and global issues in tourism. Research interests include tourism, environmental and cultural sustainability, gender studies, sense of place, and historic preservation. She has published in several journals. This is her first edited book.

Gudrun M. H. Kloes is a tour guide and operator, author, translator, and storyteller interested in interacting with nature and interpreting cultural and natural history. German born, Gudrun moved to Iceland in 1982. From 2000-2010, she was the tourism officer for SSNV (*Samtök Sveitarfélaga á Norðurlandi Vestra* - Association of Municipalities in the North West) and the main organizer behind some community initiatives in Húnaþing vestra.

Akruthai La is starting a career as an instructor in hospitality with a 10-year background in education. Akruthai is currently a Manager at the International Affairs Office of Dusit Thani College, Thailand working to ensure the college's international reputation as well as managing their expanding curriculum overseas.

Eanne P. Leite has a Master's in Tourism from the School of Arts, Sciences and Humanities (EACH) at the University of São Paulo, Brazil. Her research interests include urban tourism and migration and mobilities.

Guanhua Peng is a professional tourism planner working in tourism consulting. He obtained his Master's degree in Tourism and Sustainability at Linnaeus University in Sweden in 2018. Guanhua's interests include food tourism, ecotourism, rural community development, destination marketing, and quality control of tourist attractions.

Solène Prince is a postdoctoral researcher at the School of Business and Economics at Linnaeus University in Kalmar, Sweden. She obtained her PhD in tourism at Mid-Sweden University in 2017. Her research interests relate to rural tourism, ancestral tourism and landscape theory.

Susan L. Slocum, PhD, an associate professor at George Mason University, specializes in sustainable economic development through tourism and policy implementation. She has worked with rural communities in the US, UK, and indigenous populations in Tanzania. She has published four edited books and authored two books, including a textbook on food and agricultural tourism.

Marianna Strzelecka is a senior lecturer at the School of Business and Economics at Linnaeus University in Kalmar, Sweden. She obtained her PhD in tourism at the University of Illinois at Champaign-Urbana in 2012. Marianna's interests relate to rural communities development, protected areas tourism, environmental justice, conflicts and empowerment theory.

Acknowledgements

The editors would like to thank Sally North, Tim Goodfellow and their hard-working team at Goodfellow Publishers for their endless patience and tireless efforts. We would also like to thank all of the participants who contributed their time and experience to the studies within. Without them, this volume would not have been possible.

1 Introduction

Christina T. Cavaliere and Linda J. Ingram

The tourism system relies heavily on sustained biocultural diversity and place uniqueness. As many travellers, community members, and tourism academics have experienced, this diversity and uniqueness faces increasing global pressure and is at constant risk of annihilation and homogenisation. Overtourism, lack of or bad planning, corporate domination, economic greed, hegemony, and unequal distribution of power are forces that drive the destruction of people, places, and the planet. Humanity now recognises that we are teetering on the edge of the proverbial tipping point, and we stand in the middle of the cross-roads between conservation and mass destruction. Our human and non-human lives, wellbeing, and indeed the global tourism sector are facing the ramifications of unmanaged overuse of biocultural assets. The quintessential element of sustained existence is found in the minutiae of the intersectionality of all diversity. This diversity sustains our corporeal, spiritual, and intellectual health and is what drives us to experience the world. Yet, consumptive forces that are omnipresent and deeply imbedded in modernity are also driving the climate crisis and the perpetuation of inequalities in power and decision-making. Innovation in thought, economic structures, community resilience, and new approaches to governance are urgently needed. We hope this book serves as an exercise in questioning new approaches to these global and local imbalances.

Therefore, we invite you as the reader to explore this book and the case studies within via a vantage point of both proposed solutions and evolving contentions. We ask that you don a critical lens as you delve into the perspectives represented within this volume. The chapters are unique deconstructions of aspects of the intersection between communities and consumption. The contributing authors offer us the opportunity to understand case studies that call into question the essence of neolocalism. What is this niche trend and how is it manifesting in diverse places around the world? What are the elements of neolocalism and how can these elements serve to either slow the destruction of place or perpetuate it? Each chapter within this book offers us a distinct opportunity to question this trend and the evolving approach to tourism management, while seeking to understand the underpinnings of tourism philosophy from an applied angle. In truth, this contribution of case studies calls us, as a thought body, to question more deeply the notions of biocultural conservation, the contentions between localism and globalisation, community-based decision-making and entrepreneurship, and approaches to tourism management.

Neolocalism and the history of the term was born from the study of place. Thus, the study of sense of place is the conceptual mother of this trend and requires us to revisit the theory to situate our basic collective definition of the focus of this book. Finally, we ask you, the reader, to engage with this material with an intention to further our collective questioning of community versus corporate power and therefore, the need for economic re-appropriation. As with any approach, there are multiple angles and unavoidable aspects of duality. Within the neolocalism case studies presented in this book, we are shown subjective negative and positive results faced by the stakeholders involved. This type of critical engagement and reflective questioning of the case studies will support a growing body of thought regarding how we can collectively combat overconsumption and overtourism for the goal of conservation of biocultural diversity – the lifeblood of us all.

Revisiting sense of place: The mother of neolocalism

Flack (1997) argues the experiential angst of rootlessness (Tuan, 1980, as cited by Flack, 1997: 38) led to the generation of the cultural zeitgeist known as neolocalism in a discussion artfully connecting sense of place and neolocalism through the evolution of microbreweries. Kline et al. (2017) also link conceptualisations of the relationship between neolocalism and craft beverage tourism with various case studies showcasing local craft beverage production and consumption within two volumes (Slocum et al., 2018). While intrinsic to the introduction of this reimagining of 'local', it was left to future explorations to determine and articulate the intricacies of neolocalism. There are a wide array of viewpoints about sense of place depending upon the authors, perspectives, and disciplines (for example: Axelsson et al., 2013; Cunsolo Willox et al., 2012; Hays, 1998; Jive'n & Larkham, 2003; Jorgensen & Stedman, 2001; Pretty et al., 2003; Raymond et al., 2010; Sampson & Goodrich, 2009; Semken & Brandt, 2010; Stedman, 2002; Windsor & Mcvey, 2005). Sense of place has been characterised in a variety of ways, melding and morphing perspectives that include meaning, attachment, satisfaction, identity, and regionalisation (for example Cresswell, 2014; Klanicka et al., 2006). Defining sense of place has been problematic. Literature tends to speak of it in terms of related concepts, attributes, and associations (Kyle & Chick, 2007) and much of the research has revolved around place attachment and conceptual distinctions (Lewicka, 2011) without offering a definition. Stedman (2002) provides a concise idea of 'sense of place'; it is the meaning and value assigned to a geographic location.

Flack (1997) argues that the application of neolocalism is a way to establish sense of place, establishing their connection as embodiments of socially constructed ideas about place. Relph proposes (1976: 141), the essence of place 'lies in the largely unselfconscious intentionality that defines places as profound centers of human existence', thereby viewing place as the fusion of human and natural order; the significant center of our experience where humans begin to alter the environment to accommodate their political, economic, and social objectives creating place meaning. Meaning, however, is not inherent. Manzo (2005: 74) contends 'that it is not simply the places themselves that are significant, but rather what can be called "experience-in-place that creates meaning"'. In short, 'sense of place is a natural condition of human existence, an invariant in a changing world' (Lewicka, 2011: 209).

Placelessness is an increasingly common phenomenon. Gertrude Stein's view of placelessness was straightforward: 'when you get there, there's no there there' (Stein, 1937: 289). Essentially, placelessness occurs when the changes to a locale lead to a loss of distinctiveness and meaning (Arefi, 1999; Jackson, 1996; Relph, 1976). Early musings on sense of place have developed into an ever-evolving, ever-expanding conversation increasing in complexity (Liu & Freestone, 2016). The opposition of place and placelessness has evolved into a paradox in which the two are intertwined (Relph, 2016). When the ability/opportunity to engage with an environment is altered through loss of place (for any reason), our connection to it is damaged or lost, which can lead to feelings of rootlessness. Remedying these situations cannot be accomplished through one-size-fits all solutions. Grassroots solutions must be enacted, specific to that locale (Carolan, 2009). For example, Cavaliere and Albano (2018) examine the use of local natural and cultural imagery and folklore in craft spirit branding and its impact on craft beverage tourism in a specific geopolitical context. This empirical investigation of the use of locality within craft distilleries allows for a deeper and more nuanced understanding of the connection between local production and a sense of place. This is where neolocalism, with its focus on the local, can open the door to satisfying 'the neolocal craving' (Flack, 1997: 48) when properly applied.

Upon closer examination, there is an avenue that has largely gone unexplored in extant literature; that of the activation of the body/consciousness/senses to intertwine with place experience, thereby operationalising experience as a mediator and interpreter of both sense of place and neolocalism. Flack (1997: 48) referred to it as 'symbolic place-consciousness'. Carolan (2009: 2-4) argues 'experience and bodily practice cannot be divorced from one another. In other words, what we see (hear, smell, taste, and touch) is shaped by our doings...environment affords the body with a variety of actions and sensations…'. While at any given time, one or more senses may predominate, our understanding of our environments is the result of interaction through those senses or 'I am conscious of the world through the medium of my body…I am conscious of my body through the world' (Merleau-Ponty as cited by Carolan, 2009: 5). These understandings are informed through our own social constructions of the day-to-day world and the lived knowledge that results, meshing with Easthope (2004), Manzo (2005), Relph (1976) and Tuan (1990). Manenti (2011) takes it a step further, proposing place is foundational to interpreting coexistence thereby leading to, ultimately, the unification of human beings, highlighting the importance of a cohesive community. Neuroscience has gone so far as to identify the basic structures in the human brain that come into play when we interact with place and how those structures enhance learning, memory and recognition, and the importance of sense of place in influencing the creation of personal identity (Lengen & Kistemann, 2012).

While sense of place describes relationships between people and social-settings and human-place bonding with its rootedness, insidedness, and environmental embeddedness (Hede & Watne, 2008), neolocalism relies on an additional component for success: commercial neolocalism (Schnell, 2013). This focus on consumption is precisely what makes neolocalism vulnerable to the abuses of overconsumption by putting those with the means and power in the position to dominate local interests, shutting out those who do not have the means and power to participate, thereby crippling the triple bottom line of sustainability. In Chapter 2, Slocum and Ingram 'spotlight the evolution of sustainable tourism and situate neolocalism as the latest progression in a movement that is gaining momentum' within a framework designed to showcase neolocalism's strengths against a backdrop of the weaknesses of mass tourism. The backlash against global mass tourism and overtourism continues to increase with more and more local governments creating their own strategies and solutions in an attempt to protect fragile ecosystems, property, privacy, and cultural landmarks well as local livelihoods and lifestyles from rising costs (Gould, 2019; Henley, 2020; Hutton, 2020; Rodriguez, 2020). This chapter argues the importance of proper positioning of neolocal efforts at the local level by utilising best practices to maximise human, social, and natural capital as an alternative to the excesses of traditional mass tourism – or to avoid the same issues in destinations embracing neolocal tourism without proper planning.

Mody and Koslowsky's (2019) discussion provides a glimpse into one aspect of the excesses that can occur in neolocal tourism without holistic policy and planning when the efforts to accommodate travellers damage the very destination they seek to promote, and short-term profits are more important than long-term viability. For example, the Sedona, Arizona, USA, area's popularity as a tourist attraction is linked to neolocalism with a focus on the 'local' and (thus far) a rejection of homogeneity. However, in 2016, Arizona (the only state to do so thus far) passed the fast-tracked Senate Bill 1350 (Etienne, 2016) restricting cities and towns from regulating short term home-rentals while allowing investors to buy homes specifically for this purpose (Phillips, 2020). Hailed as 'forward thinking' and 'a win for everyone' by the Travel Technology Association (Etienne, 2016) – the results have been anything but a win for most residents. In Sedona, short term rentals comprise 30% of available rooms (Phillips, 2020) and 70% of short-term rentals are owned by people not living in the area and/or LLCs and one school has closed, in part, because newly hired teachers cannot afford housing. Some renters have been evicted so their homes can be turned into short-term rentals (Eland, 2019). In all fairness, there has been some support for the measure, but it is largely from stakeholders who own or have the resources to purchase property for short-term rentals. Arizona is not alone: issues with short-term rentals have led to restrictions and bans globally as locals grapple with concerns over loss of sense of community, property damage, and affordability (Guttentag, 2018).

Ingram explores Magnolia Market at the Silos in Chapter 3. The shopping venue has unquestionably provided a much-needed economic boon for Waco, TX, US, but there has also been a downside. Within this chapter, Ingram writes that the 'power and influence of this brand at the Silos complex is staggering. There are a variety of factors pointing to mass tourism – attendance numbers, ongoing expansion of retail space, and on-site special events – and all are indicative of large-scale commodification and consumption'. Many entrepreneurs in the surrounding area have sold their businesses, as parking became untenable, thereby decreasing economic diversity. 'By disenfranchising neighbouring business owners, minimal utilisation of local artisans, a lack of environmental controls or sustainable practices, and borderline overtourism, the Silos is indicative of what can happen to neolocalism when the financial bottom line receives greater focus than the triple bottom line'. It is difficult not to conclude that local governments have an inherent obligation to ensure that the fundamentals of neolocalism planning and policy are conducted within a multi-stakeholder framework. The runaway development and expansion of Magnolia Market at the Silos, coupled with hasty efforts by the city to keep pace with rapid growth, is an example of multiple stakeholder interests overpowered by the financial interests of a few.

Fagence, in Chapter 4, tests 'the applicability of the propositions about a new narrative of place (Schnell, 2013: 84) and nodes of geographical meaning (Ryden, 1993: 5)' by weaving together historic preservation and story-telling 'to consider the nature and scope of neolocalism from a perspective and position that is somewhat different from what may be described as "mainstream"'. Through the lens of 'tourismification', Fagence recounts the story of a nineteenth century Australian bushranger turned outlaw named Ned Kelly, reinforcing both Ryden and Schnell's hypotheses. The takeaway here is skilfully articulated by Fagence: 'The end-product can often challenge conceptualisations about "sense of place", and how it can be managed so that those fundamentals are impacted advantageously'. This is fundamental to the success of neolocalism.

In Chapter 5, Aquino and Kloes provide an Icelandic perspective of the possibilities of neolocalism – even when utilised unintentionally – 'as a strategy for place-making, cultural revitalisation, and conservation efforts of local wildlife'. This is an example of how 'community participation can lead to empowerment which leads to greater support' as they work to further regional development in the midst of the battle between small, local fishing interests and 'dominant interest groups in national politics'. This example is, arguably, what many think of as the heart of neolocalism: as an effort to not only preserve and sustain a cultural history, but to promote local economies and increase community empowerment through participation.

Unfortunately, sometimes the essence of neolocalism is distorted and then used as a justification for repression. In a 2015 commentary by a Chinese academic, neolocalism became a full blown political instrument when it was suggested advocates of Hong Kong independence would use it as a way to achieve 'full autonomy' from China as they engaged 'with anti-China and anti-communist elements in Taiwan and Southeast Asian countries to create a more deep-rooted crisis in the field of culture and ideology' (Cheung, 2015: 2-3 para). Inadvertently, this misguided commentary asserts the potential within neolocalism to enhance local pride, coalescing, empowering, and motivating communities. Neolocalism does have the potential to circumvent top down governance and corporate interests by serving as one instrument in recalibrating local governance to include equitable and inclusive decision-making from multiple stakeholders.

A compelling theme throughout the book is the capacity of the cultural festival experience to both strengthen sense of community as well as enhance understanding. Derrett provides a look into the Italian Lismore Friendship Festival in New South Wales, Australia in Chapter 6. By highlighting 'membership, influence, integration, and fulfilment of needs and shared emotional connection through sharing history and shared participation', Derrett explores the influence of neolocalism as a community builder through festival participation by recognising not only the potential power in a heritage festival's psychological underpinnings, but also by celebrating diversity both for its own sake and as a vehicle to 'grow the community's capacity to create harmony'. Derrett eloquently argues that festivals present the opportunity to support and enhance 'community resilience' as well. This broader sense of community can strengthen and enhance community engagement, and create 'new narratives in the public sphere'. The timely message of this chapter reminds us that '...my humanity is caught up and inextricably bound up in yours'. We live in a world where we are reliant not only upon each other, but where our differences serve to make us stronger when we come together and then re-emerge refreshed and revitalised to go about our daily lives. The intricate, complex, and multi-layered ways immigrants contribute to their adopted communities with the customs and practices they bring from their homelands are woven throughout this tale of Lismore during a time when globalisation has led to increasing homogenisation.

Peng, Prince and Strzelecka (Chapter 9) explore the case of the annual harvest festival on the island of Öland, in Sweden (known as Ölands Skördefest). The authors discuss this festival in light of the ways that it revives old farming traditions and how it serves to promote local cuisine, unique culinary products, and rural culture. They highlight how the festival serves to foster a sense of place through integration of various food and crafts related activities of local villages throughout the island. The authors assert that the festival results in neolocalism, as it increases local business revenue, extends the tourist season, and fosters local social sustainability. These elements discussed above demonstrate an ongoing relationship between place and culture and the ability of each to influence and alter the other.

Leite and Allis look at the links between neolocalism, multi-culturalism, and commodification in Chapter 7 by delving into the influence of Korean immigrants in the Bom Retiro district of São Paulo, Brazil. Second generation Korean immigrants have been actively renovating the district offering 'the opportunity to explore economic, urban and cultural aspects of contemporary migratory phenomena through the lens of tourism, particularly by observing "tourism mobilities"'. The authors were able to identify how second generation Brazilian-Koreans are using commodification to both assimilate and maintain their cultural distinctiveness by, for example, modifying recipes to accommodate Brazilian tastes while continuing to bring in new products from Korea. Koreans have developed a variety of local businesses and other interests that reinforce their distinct cultural identity within a multi-cultural district. Over time, these efforts extended into tourism products, primarily based in heritage and culture, as a way to interweave Korean business interests with tourism. Situations such as Bom Retiro highlight the ever-changing landscape of place, and its relationship to tourism. Rather than devolve into placelessness, these Brazilian-Koreans have moved to embrace the idea of neolocalism.

The flip side of the coin, of course, is when local interests are overshadowed by multi-national corporations, such as the public markets in Mexico City that are slowly declining or closing. These once prized *mercados* fell out of favour as Walmart, Sam's Club, Bodega Aurrerá and other venues offering cheaper, processed foods entered the marketplace. Increased gentrification and the expanding tourism market have also negatively impacted many local markets. Long considered significant to the city's cultural heritage, 'the social capital built through relationships between vendors and customers; forms of "urban popular expression," including proverbs and slang; the preservation of culinary practices; and the selling of goods that are symbolic parts of Mexican identity' (Wattenbarger, 2019).

Cavaliere et al. (Chapter 8) explore the interrelationship between heritage transportation, social capital, biocultural conservation, and sense of place through the application of the lens of neolocalism within the specific context of the Saen Saeb river canal (*khlong*) in Bangkok, Thailand. Through their analysis of issues related to both klong conservation and capacities for future development, the authors consider various issues of biocultural conservation through collaborative partnerships and good destination management strategies. Concepts involving sustainable tourism planning are interwoven with questions of equity and understandings of the distribution of both positive and negative impacts of neolocal transportation development for residents and tourists alike.

Another aspect of neolocalism focuses on consumer promotion of local interests such as the 'buy local' movement in the area of direct to consumer sales in agriculture in Chapter 10. Curtis et al. spotlight the influence of the 'locavore' movement in food tourism and how the movement, in turn, 'can support the neolocal process of crafting the tourism product around traditional agricultural practices. Specifically, dining out; visiting farmers' markets, breweries, and wineries; and participating in agritourism activities while traveling can enhance sense of place and provide enticing narratives that draw tourists to rural agricultural communities'. These efforts provide an alternative to mass agricultural production 'often governed by strict contracts and middlemen, leading to disproportionately small revenues for producers'. While the locavore movement has its detractors (for example, Desrochers & Shimizu, 2012; McWilliams, 2009), the growing popularity of the movement is evidenced by the increasing enthusiasm of a specific segment of consumers to buy local (Strolling of the Heifers, 2019) which has undoubtedly influenced the introduction of locavore grocers into the marketplace as an example. It is noted by Cavaliere (2017) that our interactions with foodscapes can perpetuate alternate ways of knowing and place consciousness. Cavaliere contends that advancing sustainability and climate consciousness can occur through engagement with place through food production and consumption.

> *'The act of eating food is the most direct energetic exchange that we engage in with the environment on a physical, psychological and, arguably, spiritual level. This direct absorption of the (un)natural world happens daily, along with the accompanying acts of cultivation, preparation and presentation of these energetic embodiments. Collectively, these food-based activities are rooted in multiple types of socio-cultural traditions, practices and performatives. Therefore, there is a significant prospect to use these acts as contemplative opportunities for alternative ways of understanding sustainable interactions within' a community or place* (Cavaliere, 2017, pg. 49).

Neolocal tourism, as noted, is well intentioned and potentially beneficial to the quadruple bottom line of environmental, economic, social, and climate change responsiveness (Fennel & Cooper, 2020). Slocum and Ingram (Chapter 2) detail the foundational underpinnings of the more familiar triple bottom line. Whereas the quadruple bottom line is expanded to include climate change as a new and separate component to consider. The authors explain that the triple bottom line of sustainability, also has the potential to devolve into more mass tourism (Mody & Koslowsky, 2019) with its excesses, or worse, as an excuse for political repression (Cheung, 2015) when proper governance is not in place. In Chapter 11, Slocum adroitly lays out 'the role of policy in the rise of neolocalism, and conversely, the inherent involvement of neolocalism in policy formation' highlighting the importance 'of community in the decision-making process'. Effective governance requires input from private and public partners working together to implement the best practices for their unique situation. Martin-Rios and Gössling (2020) write that it is time for the tourism industry to take a long, hard look at the changes taking place and become accountable to consumers on the path to the triple bottom line of sustainability. Perhaps the strength of this chapter lies in situating neolocalism squarely on that triple bottom line. With discussions about food and beverages and festivals and shopping, it is easy to relegate neolocal tourism development as just another touchy-feely concept that is timely today and soon forgotten. Instead, Slocum methodically lays out the necessity for rigorous policy and planning in neolocal tourism development as a 'valuable tool in the sustainability toolbox' for an industry muddled in excess, overtourism, and runaway growth.

Neolocalism and COVID

This book was being finalised at the very same time the COVID-19 pandemic began to impact the global community and our home communities. The authors would like to express their deep condolences to all those who have been permanently impacted through the loss of lives, income, security, and safety. The pandemic has touched all of us in so many different ways. These reflective comments are simply initial thoughts that the authors would like to explore along with the readers so that we can collectively consider the case studies presented within this book as relating to wellbeing, ethics, resilience in the face of crisis, and, ultimately, biocultural conservation. The case studies can perhaps serve to initiate deeper discussions and considerations surrounding equity and power structures within communities that provide tourism opportunities to local and foreign visitors and that, most ideally and importantly, benefit the hosts.

The COVID-19 pandemic has fundamentally changed all of our lives and in ways that we cannot fully understand at this time. However, the multiplier effect of this pandemic will undoubtedly impact lives for many generations in various ways, including health, livelihoods, community structures, access to mobility, learning environments, local and global trade, cultural norms, and ecology – among just a few. It has most certainly stopped tourism in its tracks (for a period of time) in a way that was virtually unfathomable in modernity prior to the outbreak of the virus. For most, the immediate impact of the whiplash-speed transition from hypermobility to zero mobility has caused panic and fear for many reasons – in addition to immediate health challenges, we cannot visit our families, millions cannot maintain employment locations and opportunities, engage in face-to-face hospitality or move about, even within our home communities in previously familiar ways. We do know at this juncture that there are two primary paths forward – one that involves reverting to a business-as-usual approach or, conversely, one that includes innovations for sustainable change and improved livelihoods. The immediacy of the provision of basic needs is the most crucial concern for many communities and individuals around the world. In addition, how we approach enacted change and future planning for resiliency in the face of crisis will be illuminated. There is also the larger risk of increased corporate control and homogenisation of our communities if small businesses and local entrepreneurs are not supported in their struggles to regain revenue. There is an opportunity for local through to global businesses to more deeply enact sustainable supply chain management and initiate innovations in services and sustainable structures of provision.

Thus, the central questions of good governance that supports local decision-making and stimulates conservation of uniqueness of place becomes even more crucial than before the crisis. How we collectively respond to this global pandemic will mark a new era for humanity. In addition to health and security challenges, we also have begun the process of amassing observations of the impacts of 'staying at home' – decreases in greenhouse gas emissions and air pollution, increased wild animal mobility, the use of local open space, human relationship changes, online work, governance challenges and successes, micro economies, upskilling, community engagement, and in many ways viewing the world through a new perspective. We can begin to extrapolate about the relationship between the underlying principles of neolocalism and its relationship to a new way of doing things or, at the very least, to a significant new chapter in our collective human story. How can our attention to and conservation of locality, sense of place, human and non-human relationships and social networks, entrepreneurial resilience, and at the very core – our home – support our lives?

Conclusion

Neolocalism is not intended to be a Band-Aid for the traveller seeking authenticity or for the destination seeking quick and easy socio-economic revitalisation through tourism. If the chapters in this book are indicative of anything, it is the layers upon layers of complexity within neolocalism – in planning, interpretation, implementation, and long-term viability. Even the most cursory examination reveals the difficulties in challenging a hegemonic culture of capitalism that continues to value profit for a few over the sustainability of all. When any of the components of sustainability are ignored or slighted in favour of another, the results can be detrimental to the destination – no matter the size. Neolocalism is about both participation in and resistance to the dominant culture to appropriate and re-appropriate power. In addition, it is important for all of us as consumers to learn how we can collectively strive to develop as 'critical tourism citizens' (Boluk et al., 2019) that are more aware of our ethical, social, and environmental impacts on the places we visit. It is also about the possibilities for a new type of 'growth', understanding and acceptance of diverse cultures and coming to see 'Other' as something to be celebrated and explored rather than something to be feared, hated, reviled, and mistrusted.

One of the greatest challenges to neolocal development lies in the diversity of local stakeholders; there may be differing senses of place in a locale depending upon the variety of beliefs/perspectives held by stakeholders. Recognising that sense of place is a social construct based upon relationships, jockeying for power can easily become not only an impediment to the implementation of neolocal development, but it can also lead to ongoing power struggles. Herein lies one of the greatest challenges facing neolocal development: the opportunity cost in what story/stories is/are told? This discussion has proposed the inclusion of body/consciousness/senses into sense of place and neolocalism. This brief foray was intended to be an introduction; a call to arms to begin to incorporate this important aspect of human experience into research so that, moving forward, we embrace a more holistic approach to both sense of place and neolocalism.

References

Arefi, M. (1999) Non-place and placelessness as narratives of loss: Rethinking the notion of place, *Journal of Urban Design*, **4**(2), 179-193.

Axelsson, R., Angelstam, P., Degerman, E., Teitelbaum, S., Andersson, K., Elbakidze, M. & Drotz, M.K. (2013) Social and cultural sustainability: Criteria, indicators, verifier variables for measurement and maps for visualization to support planning, *AMBIO*, **42**(2), 215-228.

Boluk, K.A., Cavaliere, C.T. & Duffy, L.N. (2019) A pedagogical framework for the development of the critical tourism citizen, *Journal of Sustainable Tourism*, **27**(7), 865-881.

Carolan, M.S. (2009) 'I do therefore there is': enlivening socio-environmental theory, *Environmental Politics*, **18**(1), 1-17.

Cavaliere, C.T. (2017) Foodscapes as alternate ways of knowing: Advancing sustainability andclimate consciousness through tactile space, in S.L. Slocum & C. Kline (eds.), *Linking Urban and Rural Tourism: Strategies for Sustainability*, Oxfordshire: CABI, pp. 49-64.

Cavaliere, C.T. & Albano, D. (2018) New Jersey craft distilleries: Sense of place and sustainability, in S.L. Slocum, C. Kline & C.T. Cavaliere (eds.), *Craft Beverages and Tourism, Volume 2*, Cham.: Palgrave Macmillan, pp. 101-117.

Cheung, G. (2015) Mainland academic rings the alarm on the rise of 'neo-localism' in Hong Kong, *South China Morning Post*, 28 December. Available at https://www.scmp.com/news/hong-kong/politics/article/1895903/mainland-academic-rings-alarm-rise-neo-localism-hong-kong. (Accessed: 13/02/2020).

Cresswell, T. (2014) *Place: An introduction*, 2nd ed, Hoboken, NJ: Wiley-Blackwell.

Cunsolo Willox, A., Harper, S.L., Ford, J.D., Landman, K., Houle, K, & Edge, V.L. (2012) "From this place and of this place:" Climate change, sense of place, and health in Nunatsiavut, Canada, *Social Science & Medicine*, **75**(3), 538-547.

Desrochers, P. & Shimizu, H. (2012) *The Locavore's Dilemma: In Praise of the 10,000-mile Diet*, New York, NY: Public Affairs.

Easthope, H. (2004) A place called home, *Housing, Theory and Society*, **21**(3), 128-138.

Eland, R. (2019) Sedona residents address short-term rentals with Arizona Rep. Bob Thorpe, *Sedona Red Rock News*, 26 July. Available at https://www.redrocknews.com/news/88888896-city-news/104464-sedona-residents-address-short-term-rentals-with-arizona-rep-bob-thorpe. (Accessed: 8/02/2020).

Etienne, S. (2016) Arizona's Governor Ducey signs SB 1350 into law, prohibiting the ban of short-term rentals, *TechCrunch*, 13 May. Available at https://techcrunch.com/2016/05/13/arizonas-governor-ducey-signs-sb-1350-into-law-prohibiting-the-ban-of-short-term-rentals/. (Accessed: 8/02/2020).

Flack, W. (1997) American microbreweries and neolocalism: "Ale-ing for a sense of place", *Journal of Cultural Geography*, **16**(2), 37-53.

Gould, L. (2019) Columbia takes aim at overtourism with unexpected alternatives, *Business Traveler USA*, 12 December. Available at https://www.businesstravelerusa.com/business-traveler-usa-story/colombias-takes-aim-at-overtourism-with-unexpected-alternatives-. (Accessed: 13/02/2020).

Guttentag, D. (2018) What Airbnb really does to a neighborhood, *BBC*, 30 August. Available at www.bbc.com/news/business-45083954. (Accessed: 12/02/2020).

Hays, R. (1998) Sense of place in developmental context, *Journal of Environmental Psychology*, **18**(1), 5-29.

Hede, A-M. & Watne, T. (2008) Leveraging the human side of the brand using a sense of place: Case studies of craft breweries, *Journal of Marketing Management*, **29**(1-2), 207-224.

Henley, J. (2020) Overtourism in Europe's historic cities sparks backlash, *The Guardian*, 25 January. Available at www.theguardian.com/world/2020/jan/25/overtourism-in-europe-historic-cities-sparks-backlash. (Accessed: 13/02/2020).

Hutton, M. (2020) Bye bye, Bangkok and Bali: why it's time to try somewhere new in Asia, *South China Post Magazine,* 12 February. Available at https://www.scmp.com/magazines/post-magazine/travel/article/3050027/bye-bye-bangkok-and-bali-why-its-time-try-somewhere. (Accessed: 13/02/2020).

Jackson, J.B. (1980) *The Necessity for Ruins,* Amherst, MA: The University of Massachusetts Press.

Jive'n, G. & Larkham P. (2003) Sense of place, authenticity and character: A commentary, *Journal of Urban Design,* **8**(1), 67-81.

Jorgensen, B.S. & Stedman, R.C. (2001) Sense of place as an attitude: Lakeshore owners attitudes toward their properties, *Journal of Environmental Psychology,* **21**(3), 233-248.

Klanicka, S., Buchecker, M., Hunziker, M. & Müller-Boker, U. (2006) Locals' and tourists' sense of place, *Mountain Research and Development,* **26**(1), 55-63.

Kline, C., Slocum, S.L. & Cavaliere, C.T. (eds.) (2017) *Craft Beverages and Tourism, Volume 1: The Rise of Breweries and Distilleries in the United States.* Cham: Palgrave Macmillan.

Kyle, G. & Chick, G. (2007) The social construction of a sense of place, *Leisure Sciences,* **29**(3), 209-225.

Lengen, C. & Kistemann, T. (2012) Sense of place and place identity: Review of neuroscientific evidence, *Health & Place,* **18**(5), 1162-1171.

Lewicka, M. (2011) Place attachment: How far have we come in the last 40 years?, *Journal of Environmental Psychology,* **31**(3), 207-230.

Liu, E. & Freestone, R. (2016) Revisiting place and placelessness, in R. Freestone & E. Liu (eds.), *Place and Placelessness Revisited,* London: Routledge, pp. 1-19.

Manenti, C. (2011) Sustainability and place identity, *Procedia Engineering,* **21**, 1104-1109.

Martin-Rios, C. & Gössling, S. (2020). Sustainable tourism: The 10 most important trends in 2020, *HospitalityNet,* 7 February. Available at https://www.hospitalitynet.org/opinion/4096895.html. (Accessed: 13/02/2020).

Manzo, L.C. (2005) For better or worse: Exploring multiple dimensions of place meaning, *Journal of Environmental Psychology,* **25**(1), 67-86.

McWilliams, J.E. (2009) *Just Food: Where Locavores Get It Wrong and How We Can Truly Eat Responsibly,* New York, NY: Little, Brown.

Mody, M. & Koslowsky, K. (2019) Panacea or peril? The implications of Neolocalism as a more intrusive form of tourism, *Boston Hospitality Review,* **7**(1).

Phillips, M. (2020) Sedona community asks state legislature to change state's short term rental laws, *ABC15.com,* 10 January. Available at https://www.abc15.com/news/region-northern-az/sedona/sedona-community-asks-legislature-to-change-states-short-term-rental-laws. (Accessed: 8/02/2020).

Pretty, G.H., Chipuer, H.M. & Bramston, P. (2003) Sense of place among adolescents and adults in two rural Australian towns: The discriminating features of place attachment, sense of community and place dependence in relation to place identity, *Journal of Environmental Psychology,* **23**(3), 273-287.

Raymond, C.M., Brown, G. & Weber, D. (2010) The measurement of place attachment: Personal, community and environmental connections, *Journal of Environmental Psychology,* **30**(4), 422-434.

Relph, E. (1976) *Place and Placelessness,* London: Pion.

Relph, E. (2016) Senses of place and emerging social and environmental challenges, in J. Eyles & A. Williams (eds.), *Sense of Place, Health and Quality of Life,* London: Routledge, pp. 31-44.

Rodriguez, C. (2020) Should you visit these beautiful European destinations suffering from overtourism?, *Forbes,* 30 January. Available at https://www.forbes.com/sites/ceciliarodriguez /2020/01/30/to-go-or-not-to-go-to-these-beautiful-european-places/#4e916f7029e4. (Accessed: 13/02/2020).

Ryden, K. (1993) *Mapping the Invisible Landscape: Folklore, Writing, and the Sense of Place,* Iowa City, IA: University of Iowa Press.

Sampson, K.A. & Goodrich, C.G. (2009) Making place: Identity construction and community formation through "sense of place" in Westland, New Zealand, *Society & Natural Resources,* **22**(10), 901-915.

Schnell, S. (2013) Deliberate identities: becoming local in America in a global age, *Journal of Cultural Geography,* **30**(1), 55-89.

Semken, S. & Brandt, E. (2010) Implications of sense of place and place-based education for ecological integrity and cultural sustainability in diverse places, in D. Tippins, M. Mueller, M. van Eijick & J. Adams (eds.), *Cultural Studies and Environmentalism: The Confluence of EcoJustice, Place-Based (Science) Education, and Indigenous Knowledge Systems,* New York, NY: Springer, pp. 287-302.

Slocum, S.L., Kline, C. & Cavaliere, C.T. (eds.). (2018) *Craft Beverages and Tourism, Volume 2*. Cham: Palgrave Macmillan.

Stedman, R.C. (2002) Toward a social psychology of place: Predicting behavior from place-based cognitions, attitude and identity, *Environment & Behavior,* **34**(5), 561-581.

Stein, G. (1937) *Everybody's Autobiography,* New York, NY: Random House.

Strolling of the Heifers (2019), 'Agriculture census shakes up the Locavore index: Vermont still on top; California jumps to second place'. Available at https://www.strollingoftheheifers.com/locavore/. (Accessed: 12/02/2020).

Tuan, Y-F. (1980) Rootedness vs sense of place, *Landscape,* **24**, 3-8.

Tuan, Y-F. (1990) *Topophilia: Study of Environmental Perceptions, Attitudes, and Values,* New York, NY: Columbia University Press.

Wattenbarger, M. (2019) Are we killing the markets of Mexico City?, *Eater,* 9 December. Available at www.eater.com/2019/12/9/20963654/mexico-city-public-markets-in-trouble-over-tourism-gentrification. (Accessed: 14/02/2020).

Windsor, J.E. & Mcvey, J.A. (2005) Annihilation of both place and sense of place: the experience of the Cheslatta T'En Canadian First Nation within the context of large-scale environmental projects, *Geographical Journal,* **171**(2), 146-165.

2 Sustainability and Neolocalism: Complementary Ideologies

Susan L. Slocum and Linda J. Ingram

Abstract

This chapter introduces neolocalism as a valuable tool in the sustainability toolbox. Neolocalism, often considered the opposition to globalisation, has the potential to support ecologically and culturally appropriate tourism development grounded in place attachment. While slow progress is being made by tourism industries, ethical change can be seen at local levels, embraced and enhanced by entrepreneurs and destinations that value their unique attributes. This chapter intends to navigate these changes as a means to spotlight the evolution of sustainable tourism and situate neolocalism as the latest progression in a movement that is gaining momentum. By reviewing a brief history of sustainability thought, and explaining the role of the experiential, knowledge-based, and creative economies, this chapter encourages optimism for tourism destinations. As sustainable tourism evolves and scholars acknowledge the emergence of sustainable mass tourism, there is pressure for destinations to move beyond the green growth approach to tourism. The adoption of neolocalism could facilitate exciting opportunities that support the three pillars of sustainability.

Introduction

While globalisation has contributed to the growth of tourism, overtourism has contributed to the loss of place distinctiveness and identity (Ginting et al., 2017), in addition to environmental degradation, habitat fragmentation and negative socio-cultural impacts (Choi & Sirakaya, 2006). As awareness of the inequities, damage, and unsustainability from mass tourism increases, more and more tourists seek to experience travel destinations in new ways. These trends have led to the introduction of neolocalism into tourism development as travellers aspire to immerse themselves in the 'distinctively local' in an effort to have an 'authentic' experience. Rather than have their vacation shaped by mass tourism enclaves and staged experiences controlled by tourism providers (Mody & Koslowsky, 2019), while simultaneously avoiding the negative impacts associated with mass tourism (Archer et al., 2005), many tourists are seeking authentic, local, and placed-based activities that provide holistic connections to the places they visit. It is not surprising that destinations are moving to neolocalism as a way to embrace these consumer trends. This transition has fostered new partnerships, enhanced the tourism offer, and instilled a new sense of local ownership over the identity and authenticity of tourism products, messages, and experiences (Litzky et al., 2016).

This chapter introduces neolocalism as a valuable tool in the sustainability toolbox. Neolocalism, often considered the opposition to globalisation, has the potential to support ecologically and culturally appropriate tourism development grounded in place attachment (Wright & Eaton, 2018). One of the strengths of the neolocal movement lies in its focus on reclaiming and revitalising the uniqueness of a tradition, craft, or the sense of place of a locality. By utilising creative talents that embrace cultural and environmental aspects of community, neolocalism provides a valuable lens through which to reflect on sustainable tourism to date.

While Higgins-Desbiolles (2010) expresses frustration with the slow progress being made by tourism industries, ethical change can be seen at local levels, embraced and enhanced by entrepreneurs and destinations that value their unique attributes. This chapter intends to navigate these changes as a means to spotlight the evolution of sustainable tourism and situate neolocalism as the latest progression in a movement that is gaining momentum. By reviewing a brief history of sustainability thought, and explaining the role of the experiential, knowledge-based, and creative economies, this chapter argues that all hope is not lost. Ceccarelli (2019) contends that through collaboration and innovation, neolocalism can encourage critical debate and find alternatives that address current weakness in traditional mass tourism development.

This chapter begins with an overview of sustainability as a means to frame the debate surrounding the negative influences of modern tourism. Using an historical narrative reminds readers that sustainability is both an old concern and a new paradigm, one that is ever changing and advancing. Even as the United Nations moves from the Millennium Goals to the Sustainable Development Goals, emphasis is placed on appropriate levels of development that can improve the quality of life for residents, yet conserve traditional resources, and include new ideas, perceptions, and alternative development frameworks. Cultural tourism, one of the fastest growing components of tourism (United Nations World Tourism Organisation, 2018), supports creative economies and product development, which in turn assists in the promotion of neolocal development and marketing (Holtkamp et al., 2016). Richards (2009) argues that neolocalism may, in fact, put the power of commodification into the hands of the very people who value authenticity, allowing for local construction of the inherent value and meaning of authenticity as communities change, evolve, and negotiate their identity over time. Yet, as we face challenges to sustainability, this chapter argues that neolocalism can support social and human capital development, which has the potential to support sustainability.

As our knowledge grows, so does our ability to facilitate change. By way of a new approach to sustainable tourism, there is much still unknown about the consequences and pitfalls of neolocal development. However, this chapter hopes to situate neolocalism in the ongoing discussion surrounding future development paths and appropriate development patterns. While tourism has many inherently negative consequences, a wanderlust for the world will always provide economic opportunities to exploit the less fortunate. Therefore, neolocalism seems well poised to put power into local hands and encourage greater participation in the decision-making process.

History of sustainability: thought and theory

The word 'sustainability' is relatively new. The first book to include concrete perceptions of sustainability was entitled *Blueprint for Survival* published in 1972. The book was written by the editorial board of *The Ecologist*, specifically Goldsmith and Allen of the Conservation Society, and signed by over 30 prominent scientists of the time. The premise was a return to small, tribal, pre-industrial societies as a means to control population growth, reduce pollution, increase social cohesion, and prevent the 'the breakdown of society and the irreversible disruption of the life-support systems on this planet' (The Ecologist, 1972: 20). However, the rapid rise in the use of sustainability in the 1980s, culminating in the 1987 United Nation's World Commission on Environment and Development (the Brundtland Commission), has brought the term into the everyday vernacular. Today, the search term 'sustainability' can produce over 150 million google hits (Caradonna, 2014).

While the word may be contemporary, the concept is not. From the onset of early civilisation, intensive agriculture and over-use of natural resources often led to the decline of civilisations (Diamond, 2005). Examples may include the Maya, Anasazi, and Easter Island communities. The Industrial Revolution increased population growth rates, influencing scholars, such as Thomas Malthus and John Stewart Mills, to propose alternative forms of society. Many of the scholars of the 1800s formed the new scientific disciplines of political ecology, environmental economics, and environmental science. Even at its infancy, thoughts on sustainability included both a human and environmental element, as their interconnectedness could not be ignored. The conundrum that poverty destroys the environment, yet the decline of the environment increases poverty, is as true today as in Max Weber and Karl Marx's time. Capitalism, or the ability of a few to capitalise on common pool resources at the expense of others, forms the foundation of political sociology and sustainability (Lockie, 2015).

Yet, the social movements that propelled sustainability concepts toward the mainstream occurred outside academia, primarily in a renewed interest in nature, human health, and a sense of spiritually expressed in the writings of Wordsworth, Emerson, Thoreau, and Muir. Ideas of Transcendentalism, started by Emerson and his peers in the 1830s, and furthered by Muir, had a profound impact on the formation of the US National Park Service and the preservation of ecosystems for human enjoyment, relaxation, and godly experiences (Hall, 2002). Conservation initiatives increased in the early 20th century and reached the highest international institution in 1949 when the United Nations convened The U.N. Scientific Conference on the Conservation and Utilization of Resources.

During the prolonged economic expansion that occurred after World War II, scholars recognised that the massive consumption of resources could not be maintained over the long term, and shortages would eventually occur. However, the common belief, based on the work of David Ricardo in 1951, was that once a product or a factor input became scarce, new technologies would be developed to replace the scarce input (Du Pisani, 2006). This philosophy justified the continued economic colonisation and resource depletion occurring in poor parts of the world, which supported mass consumerism in wealthier countries.

The new social era of the 1960s and 1970s promoted a more ecologically sound and humanitarian form of development. Concerns ranged from social inclusion, including black, women, and gay rights, to environmental concerns related to pesticide use in food, pollution, and endangered species. *The Silent Spring*, written by Rachel Carson, had a direct impact on the rising Green Movement, leading to the first Earth Day in 1970. As media coverage publicised the 'imminent ecological crisis' (Du Pisani, 2006: 89), Greenpeace and Friends of the Earth became more visible, and green political parties emerged in many developed countries, paving the way for sustainable development and a new paradigm for industry.

Sustainability is founded on a premise that current generations have a responsibility to future generations by ensuring they have the same opportunities to enjoy natural and cultural resources as a means to ensure their continued survival and quality of life (Keeble, 1988). Sustainable development is both controversial and contentious as many political adversaries believe sustaining cultures and environments is counter-intuitive to growing an economy and meeting the needs of current populations. However, one vital aspect to sustainability is the use of sound scientific methods to support the appropriate types and levels of development to avoid exploitation and over use of resources (Jafari, 1990). Moreover, the acceleration of species extinction, cultural assimilation, and climate change is providing the documentation needed to bring governments and business in line with sustainability consciousness.

Sustainability hinges on three specific pillars, also known as the triple bottom line: the economy, the environment, and society. Gössling and Hall (2013) divide production and consumption into four categories.

- The business as usual approach results in economic exchanges based on mass commodities and places special emphasis on economic returns. One could argue this approach is laissez-faire and not a part of sustainable development.
- The green growth approach encourages greater efficiency through green economic growth and a reliance on technological and market solutions to balance the economy, environment, and society. Some scholars refer to this as weak sustainability because it focuses on the reduction of resource use in order to produce goods and services at a cost savings (Neumayer, 2003).
- The traditional sustainable approach is based on traditional methods of production, where emphasis is placed on small-scale business development that is culturally and ecologically appropriate to serve local needs and promote opportunities for the poor.
- The steady-state approach promotes a balance between appropriate production and consumption levels with environmental and cultural conservation. The traditional sustainable approach and the steady-state approach align with forms of strong sustainability where environmental conservation and social progress are prioritised over economic growth (Neumayer, 2003).

In 2000, the international community committed to sustainable development through the adoption of the Millennial Development Goals by the United Nations. These ambitious goals were revised in 2015 as the Sustainable Development Goals in an effort to put pressure on policy organisations to prioritise sustainability and to recognise the needs of communities and the environment first in future development initiatives (see Table 2.1). As sustainability evolves, new elements are included, such as infrastructure, access to clean water, reliable energy, and inclusive economic growth, in an attempt to recognise the interdependence between and within the triple bottom line.

Table 2.1: Millennium and Sustainable Development Goals

Millennium Development Goals (2000) To be achieved by 2015
Eradicate extreme poverty and hunger
Achieve universal primary education
Promote gender equality and empower women
Reduce child mortality
Improve maternal health
Combat HIV/AIDS, malaria, and other diseases
Ensure environmental sustainability
To develop a global partnership for development
Sustainable Development Goals (2016): To be achieved by 2030
End poverty in all its forms everywhere
End hunger, achieve food security and improved nutrition and promote sustainable agriculture
Ensure healthy lives and promote well-being for all at all ages
Ensure inclusive and equitable quality education and promote lifelong learning opportunities for all
Achieve gender equality and empower all women and girls
Ensure availability and sustainable management of water and sanitation for all
Ensure access to affordable, reliable, sustainable and modern energy for all
Promote sustained, inclusive and sustainable economic growth, full and productive employment and decent work for all
Build resilient infrastructure, promote inclusive and sustainable industrialisation and foster innovation
Reduce income inequality within and among countries
Make cities and human settlements inclusive, safe, resilient and sustainable
Ensure sustainable consumption and production patterns
Take urgent action to combat climate change and its impacts by regulating emissions and promoting developments in renewable energy
Conserve and sustainably use the oceans, seas and marine resources for sustainable development
Protect, restore and promote sustainable use of terrestrial ecosystems, sustainably manage forests, combat desertification, and halt and reverse land degradation and halt biodiversity loss
Promote peaceful and inclusive societies for sustainable development, provide access to justice for all and build effective, accountable and inclusive institutions at all levels
Strengthen the means of implementation and revitalise the global partnership for sustainable development

Source: United Nations (2015)

From sustainable tourism to creative tourism – a new paradigm

Sustainable tourism is a subset of sustainable development and addresses the experiential nature of travel. Specifically, as an export industry, visitors travel to the places of production, further aggravating the negative impacts associated with the industry (Chang, 1999), such as excessive resource consumption (water, energy and waste), commodification of culture, and the promotion of western values. Tourism also provides avenues to enhance destinations as tourists expect certain amenities, which may become available and accessible to local residents, and, increasingly, seek out authentic experiences dependent on pristine environments and traditional cultures. Therefore, tourism can provide the economic impetus to fund conservation or poverty alleviation initiatives (Goodwin, 1995). The tourism sector is dependent on the quality of destination attributes, which often include the resources inherent in the triple bottom line of sustainability (Slocum & Curtis, 2018).

Sustainable tourism is well researched and has become a prominent paradigm for tourism development. However, strategies that support the triple bottom line of sustainability have been challenging to implement. Some of the hurdles include apathy by local communities to engage in the political or developmental process of tourism, a lack of expertise on sustainable practices, and a lack of fiscal resources to convert traditional modes of tourism production (Tosun, 2000). New philosophies aligned with ecotourism have included appropriate levels and types of tourism as a means to ensure a vibrant economy without jeopardising local resources (Weaver & Lawton, 2014). Moreover, cultural tourism provides valuable opportunities to promote tourism, showcases local traditions or historical narratives, and instils a sense of pride that can lead to greater involvement by community members (McKercher et al., 2004)

Culture has become a viable path to enhance tourism destinations and create economic opportunities through new attractions, cultural routes, and heritage centres (Richards, 2009). Cultural tourism accounts for approximately 37% of global tourism, and could grow upwards of 15% yearly (United Nations World Tourism Organisation, 2018). Moreover, cultural practices and artisan production are essential to the consumption of culture. Richards (2003) writes, 'There is a general perception that cultural tourism is "good" tourism that attracts high spending visitors and does little damage to the environment or local culture while contributing a great deal to the economy and support of culture' (p. 1). Cultural tourism occurs when participation in a cultural or heritage activity is a significant motivator for traveling and includes both tangible culture, such as museums and cultural centres, and intangible culture, including performing and visual arts, crafts, and festivals.

In order for cultural tourism to be a form of sustainable tourism, it should meet certain thresholds. The Canadian Federal Provincial Territorial Ministers of Culture and Heritage (2012) highlight three primary principles:

- It involves a quality visitor encounter with a cultural experience that is authentic, and that is not created solely for the purpose of attracting tourists;
- It must ensure that the quality and integrity of the cultural resource/experience are conserved; and
- It is based on a mutually beneficial partnership between the cultural tourism sectors, and must be a revenue generator for all sectors.

Cultural tourism has the potential to support both the traditional sustainable approach and/or the steady state approach to sustainability when emphasis is placed on small-scale artisan production that enhances the visitor experience (Gössling & Hall, 2013). Because tourism is experiential in nature and is produced and consumed simultaneously (Pantzar & Shove, 2005), tourism has become a driver of the creative economy (Richards, 2009). Specifically, it is the creative process inherent in the production of cultural tourism products that enhances and supports traditional industries and appeals to tourism consumers. As mass tourism continues to rely on staged experiences for mass consumption, the tourism product and the destination are requiring new forms of innovation as a means of differentiation. The creative economy is fast becoming a force of sustainability.

Neolocalism – a tool in the sustainability toolbox

Neolocalism was first suggested by Flack (1997) who noted 'Americans are more likely to long for and seek out unique places' (p. 38). Shortridge (1996) understood neolocalism as 'a deliberate seeking out of regional lore and local attachment by residents (new and old) as a delayed reaction to the destruction in modern America of traditional bonds to community and family' (p. 10). Neolocalism is, therefore, a form of resistance to the ongoing homogenisation of place and culture (Wright & Eaton, 2018). However, neolocalism is not restricted to America; it has become a global phenomenon in response to the negative impacts of globalisation and corporatisation (see for example: Das, 2005; Parnwell, 2007; Rogerson & Collins, 2015; Schnell, 2013).

The concept of 'local' has come to be considered the antithesis of 'global'. Local implies a rootedness to community, transparency of production, non-corporate ownership, embracing sustainability, and a place to experience authenticity (Schnell, 2011). The focus on local and what it can provide has influenced, for example, the rise of the craft beer industry (Kline et al., 2017), local food industries (Schnell, 2011), agritourism (Wright & Eaton, 2018), and homestay/vacation rental companies (Mody & Koslowsky, 2019) through the symbolisation of local products as a way to enhance community identity (Adema, 2006). This focus on local means of production and community responsibility contrasts corporatised tourism where the tenets of capitalist globalisation and trans-national corporations manipulate tourism for profit maximisation, most of which is lost to local communities through leakage and the 'exploitation and commodification of all factors of production including people, cultures and environments' (Higgins-Desbiolles, 2010: 120). Given the sluggishness of the tourism industry when it comes to implementing sustainability, the maximisation of profits through exploitation is indicative of the lack of concern most tourism providers have for the interests of local stakeholders and environments (Higgins-Desbiolles, 2010). Consequently, the commodification of local products and resources, when combined with mass tourism, can result in environmental degradation, loss of distinctiveness, economic leakage, and a variety of inequities. For example, while food and drink have traditionally been a part of the tourist experience, it has been argued that it is the kind of food and drink, where it is sourced, and how it is consumed, which has long-term significance for local environmental, cultural, and economic interests (Sims, 2009).

The links between locally sourced food and drink, locally produced goods and services, and sustainability are well documented. A growing body of research indicates that locally sourced food and drink have the potential to not only initiate a multiplier effect in an area, they can also contribute to the distinctiveness and 'authenticity' travellers seek while enhancing the triple bottom line of sustainability. Gastronomic distinctiveness, combined with a variety of locally produced goods and services, are necessary components for a tourism destination beneficial to the interests of both tourists and locals. This link between food and local can symbolise the authenticity of place and is key to sustainable tourism development while influencing local cultural and social legacies (Gálvez et al., 2017). While authenticity remains a contested topic among researchers, consumers continue to seek it out as a means to connect with a world from which they feel increasingly disconnected (Sims, 2009).

Neolocalism and sustainability – authenticity for consumption

Debates about the connection between authenticity and commodification in tourism and their influence on communities began with MacCannell's *The Tourist* in 1976. While the relationship between tourist consumption and local can embody authenticity (Gálvez et al., 2017), there is also an argument to be made that tourism alters authenticity through the commodification of culture, leading to overconsumption of heritage and place by tourists (MacLeod, 2006). In other words, commodification of local products can be understood as the consumption of place and the consumption of identity (Higgins-Debiolles, 2010) for economic gain often without consideration for the environment or local culture. Authenticity is damaged or lost – for both the community and visitors – when locally developed products and performances are redesigned for the tourism market (MacLeod, 2006). As consumption in contemporary societies is an expression of individuality and independence (Higgins-Debiolles, 2010), commodification of products and performances by tourism providers then becomes crucial for de rigueur consumption by tourists. This, in turn, leads to exploitation as visitor numbers increase. Local products, ranging from food and beverages to local history and crafts, can be commodified to communicate ideas about cultural characteristics or a tourism product's authenticity or sustainability (Frisvoll et al., 2016). The challenge lies in developing tourism policies and plans that will allow for sustainable commodification and consumption of local products and services.

Tourism trends often reflect those seen in society. The shift in consumer preferences from tourism enclaves and staged authenticity to immersive, meaningful travel experiences means tourism providers have to generate products that can satisfy the changing needs of contemporary travellers who are also increasingly aware of the negative impacts of tourism. Neolocalism, with its non-corporate, community building, small scale production at the local level and artisan promoting culture, is intrinsic to the idea of sustainable tourism development (Mody & Koslowsky, 2019).

Cultural tourism, in all its forms, is an established method used to attract visitors and, as noted, can also be a vehicle for supporting sustainable tourism when attention is focused on small-scale artisan production. Festivals, food and beverages, and locally generated artisanal products can all be considered under the umbrellas of both cultural tourism and neolocal tourism as part of the creative economy. It has been argued that the influence of the creative economy is so powerful it will not only be key to economic development in the twenty-first century, it could be the dominant form of tourism development. By 2002, creative economies in OECD (Organisation for Economic Co-operation and Development) countries experienced growth at twice the rate of service industries and four times that of manufacturing (Tepper, 2002).

Challenges to neolocalism and sustainability – a path forward?

Sustainable tourism based on the concept of neolocalism is not assured. Sustainable tourism remains a contested topic subject to varying interpretations by tourism providers, academics, researchers, and travellers. The review of sustainable tourism research by Bramwell et al. (2017) highlights the evolving nature of the field. The idea of 'sustainability' is one used by many tourism developers and providers for its promotional value with tourists. Use of the word 'sustainable' is included in many forms of development with little to no understanding of its meaning or its application in planning or management. Sustainable tourism development is the drumbeat used to introduce tourism to un- or underdeveloped areas as external agents take over valuable lands and cultural resources while disparaging local populations' use, cultures, and practices (Cohen, 2002). Despite over forty years of research, a unified definition for sustainability remains elusive (Caradonna, 2014).

The most holistic sustainable tourism development policies and programs are for naught without environmentally friendly behaviours from tourists. Research indicates, overall, individuals reduce their environmentally friendly behaviours when traveling (Slocum & Curtis, 2015). Most tourists desire a break from their day-to-day efforts and/or believe it is up to the government to regulate tourism destinations (López-Sánchez & Pulido-Fernández, 2016). Tourists are also often unaware of the specifics of environmentally friendly practices when traveling. The tourism industry has, by and large, not only failed to implement sustainable tourism policies, but also failed to influence and educate tourists by encouraging pro-environmental and culturally appropriate behaviours (Dolnicar, 2015).

Perhaps one of the biggest challenges to effective implementation of sustainability in any neolocal tourism destination is the possibility that as a destination increases in popularity so will the desire for expansion. The tourism industry is well known for ignoring imposed limits to tourist numbers, carrying capacities, infrastructure development, and other impediments to growth and profit (Higgins-Debiolles, 2010). Local stakeholders also must be willing to sacrifice short-term profits for long-term sustainability in an effort to sustain the existence of tourism to a destination.

The key difference between neolocalism and localism is in the intentional construction of cultural and place-based narratives. According to Holtkamp et al. (2016), neolocalism requires a conscious effort by businesses to foster a sense of place based on the attributes of their community and includes three primary indicators: (1) use of local names and images in labelling and marketing; (2) environmental sustainability; and (3) social and community engagement. In turn, these indicators must contribute to identity formation and sustainable practices if they are to lead to self-reliant communities. Neolocalism is a model that is open, adaptable, and responsive. This, in turn, provides avenues for the 're-evaluation of local values and traditional knowledge in the belief that if re-interpreted through new, open, small technologies, through sharing, collaborating, and innovating, these approaches and expertise can help find solutions to contemporary problems and needs' (Ceccarelli, 2019: 935). Convincing stakeholders to develop the appropriate type and size of the tourism industry requires the development of social capital and a recognition that cooperation and 'coopetition' ensures the long-term viability of all key players.

Therefore, neolocalism requires human and social capital if it is to provide the impetus to develop locally appropriate destinations. Human capital generally describes the knowledge, skills, health, and self-esteem needed to build talents, increase productivity, and participate in the economic system (Stone & Nyaupane, 2016). Conversely, social capital includes the trust and reciprocity that allows for relationship building and collaboration that can lead to the achievement of both private and collective goals (Bhandari & Yasunobu, 2009). Social capital and human capital are intertwined, in that 'increased social capital can provide an avenue to create innovative local tourism products… (resulting in) groups of businesses that complement each other and provide a variety of experiences for the tourists' (Slocum & Klitsounova, 2019: 75). Artisan communities generally possess strong social capital and are willing to share the knowledge needed to build human capital (Lourens, 2007; Slocum, 2015). Transferring these resources to tourism industries can encourage the co-construction of place-based tourism, positively influencing sustainable tourism development around authentic destination attributes. As an avenue to counter some of the barriers to sustainable tourism development (Tosun, 2000), neolocalism, as a tool for sustainability, could potentially breakdown barriers to human and social capital accumulation and encourage a more participatory approach to place-based tourism.

Conclusion

This chapter has argued that neolocalism provides exciting potential to embrace a more sustainable tourism industry at the local level through cultural resources, creative economies, and authenticity. As a relatively new approach to sustainability, many of the challenges, opportunities, and consequences are yet unknown, nonetheless there is ample evidence to suggest that neolocalism is worthy of further exploration. As sustainable tourism evolves and scholars acknowledge the emergence of sustainable mass tourism (Weaver, 2012), there is pressure for destinations to move beyond the green growth approach to tourism (Gössling & Hall, 2013). The adoption of neolocalism could facilitate exciting opportunities that support the three pillars of sustainability.

To reinforce the pursuit for appropriate levels and types of tourism development, there is a need to place the power of production into the hands of those most influenced by unsustainable forms of tourism. Therefore, we encourage future research into the growing neolocal phenomenon. There is a lack of understanding about how neolocalism influences local attributes, including authenticity, commodification, and ongoing consumption of natural and cultural resources. As a first step, analysing the impacts of neolocalism should guide further discussions as a means to inform critical debate (Ceccarelli, 2019). Moreover, the collection and dissemination of best-practice for both creative entrepreneurs and destinations can better guide development decisions and participatory approaches to policy formation. Lastly, to date, neolocalism has resided within the food and drink research arena. There are obvious connections to other aspects of the creative economy, including intangible heritage, such as event planning, artisan industries, and place-based image creation. However, there potentially could be larger and further-reaching uses for neolocally constructed tourism narratives, both within urban and rural tourism marketing, entrepreneurship, and destination loyalty research. There is a call for a more creative approach to measuring, defining, and understanding sustainability as we move from its introduction into our everyday vernacular towards a society keenly aware of the impacts of unsustainable decisions. We hope scholars critically analyse neolocalism as a potential tool towards a more sustainable form of experiential consumption.

References

Adema, P. (2006) *Festive foodscapes: iconizing food and the shaping of identity and place*, PhD dissertation, University of Texas, Austin.

Archer, B., Cooper, C. & Ruhanen, L. (2005) The positive and negative impacts of tourism, in W.F. Theobald (ed.), *Global Tourism*, 3rd edn, Burlington, MA: Elsevier pp. 79-102.

Bhandari, H. & Yasunobu, K. (2009) What is social capital? A comprehensive review of the concept, *Asian Journal of Social Science*, **37**(3), 480–510.

Bramwell, B., Higham, J., Lane, B. & Miller, G. (2017) Twenty-five years of sustainable tourism and the Journal of Sustainable Tourism: Looking back and moving forward, *Journal of Sustainable Tourism*, **25**(1).

Caradonna, J.L. (2014) *Sustainability: A history*. New York, NY: Oxford University Press.

Ceccarelli, N. (2019) Neo-Local design. Looking at 'our local contexts' as potential resources, *The Design Journal*, **22**(1), 31-946,

Chang, T.C. (1999) Local uniqueness in the global village: heritage tourism in Singapore, *Professional Geographer*, **51**(1), 91-103.

Choi, C. & Sirakaya, E. (2006) Sustainability indicators for managing community tourism, *Tourism Management*, **27**(6), 1274-1289.

Cohen, E. (2002) Authenticity, equity and sustainability in tourism, *Journal of Sustainable Tourism*, **10**(4), 267-276.

Das, K. (2005) *Micro and Small Enterprises*, Alternative Economic Survey, India, pp. 111-120.

Provincial Territorial Ministers of Culture and Heritage Canada. (2012) *Cultural and Heritage Tourism: A Handbook for Community Champions*.

Diamond, J. (2005) *Collapse: How societies choose to fail or succeed*, New York, NY: Viking Books.

Dolnicar, S. (2015) Environmentally sustainable tourists?, in C.M. Hall, S. Gössling & D. Scott (eds.), *The Routledge Handbook of Tourism and Sustainability*, New York, NY: Routledge, pp. 140-150.

Du Pisani, J.A. (2006) Sustainable development–historical roots of the concept, *Environmental Sciences*, **3**(2), 83-96.

The Ecologist. (1972) *A Blueprint for Survival*, Westminster: Penguin Books.

Frisvoll, S., Forbord, M. & Blekesaune, A. (2016) An empirical investigation of tourists' consumption of local food in rural tourism, *Scandinavian Journal of Hospitality and Tourism*, **16**(1), 76-93.

Gálvez, J.C.P., Granda, M.J., López-Guzmán, T. & Coronel, J.R. (2017) Local gastronomy, culture and tourism sustainable cities: The behavior of the American tourist, *Sustainable Cities and Society*, **32**, 604-612.

Ginting, N., Nasution, A.D. & Rahman, N.V. (2017) More attractive more identified: distinctiveness in embedding place identity, *Procedia Environmental Sciences*, **37**, 408-419.

Goodwin, H. (1995) Tourism and the environment, *Biologist*, **42**(3), 129-133.

Gössling, S. & Hall, C.M. (2013) Sustainable culinary systems: An introduction, in C.M. Hall & S. Gössling (eds.), *Sustainable Culinary Systems: Local Food, Innovation, Tourism, and Hospitality*, London: Routledge, pp. 3–44.

Hall, C.M. (2002) The changing cultural geography of the frontier: National parks and wilderness as frontier remnant, in S. Krakover & Y. Gradus (eds.), *Tourism in Frontier Areas*, Maryland: Lexington Books, pp. 283-98.

Higgins-Desbiolles, F. (2010) The elusiveness of sustainability in tourism: The culture-ideology of consumerism and its implications, *Tourism and Hospitality Research*, **10**(2), 116-129.

Holtkamp, C., Shelton, T., Daly, G., Hiner, C.C. & Hagelman III, R.R. (2016) Assessing neolocalism in microbreweries, *Papers in Applied Geography*, **2**(1), 66-78.

Jafari, J. (1990) Research and scholarship: The basis of tourism education, *Journal of Tourism Studies*, **1**(1), 33-41.

Keeble, B.R. (1988) The Brundtland report: 'Our common future', *Medicine and War*, **4**(1), 17-25.

Kline, C., Slocum, S.L. & Cavaliere, C.T. (eds.) (2017) *Craft Beverages and Tourism, Volume 1: The Rise of Breweries and Distilleries in the United States*, New York, NY: Springer.

Litzky, B., Andersson, L. & Smith, W.P. (2016) Stay with me: an investigation into the durability of social capital accumulation for urban entrepreneurs, in *Proceedings of the 27th Annual Conference of the International Association for Business and Society*, 16-19 June, Park City, Utah, (Vol. 27, pp. 146-157).

Lourens, M. (2007) Route tourism: A roadmap for successful destinations and local economic development, *Development Southern Africa*, **24**(3), 475–490.

Lockie, S. (2015) Why environmental sociology? *Environmental Sociology*, **1**(1), 1-3.

López-Sánchez, Y. & Pulido-Fernández, J.I. (2016) In search of the pro-sustainable tourist: A segmentation based on the tourist 'sustainable intelligence', *Tourism Management Perspectives*, **17**, 59-71.

MacLeod, N. (2006) Cultural tourism: Aspects of authenticity and commodification, in M.K. Smith & M. Robinson (eds.), *Cultural Tourism in a Changing world: Politics, Participation and (Re) presentation*, Clevedon: Channel View Publications, pp. 177-190.

McKercher, B., Ho, P.S. & du Cros, H. (2004) Attributes of popular cultural attractions in Hong Kong, *Annals of Tourism Research*, **31**(2), 393-407.

Mody, M. & Koslowsky, K. (2019) Panacea or peril? The implications of neolocalism as a more intrusive form of tourism, *Boston Hospitality Review*, **7**(1).

Neumayer, E. (2003) *Weak Versus Strong Sustainability: Exploring the limits of two opposing paradigms*, London: Elgar.

Pantzar, M. & Shove, E. (2005) *Manufacturing Leisure. Innovations in happiness, wellbeing and fun*, Helsinki: National Consumer Research Centre.

Parnwell, M.J. (2007) Neolocalism and renascent social capital in northeast Thailand, *Environment and Planning D: Society and Space*, **25**(6), 990-1014.

Richards, G. (2003) What is Cultural Tourism? in A. van Maaren (ed.) *Erfgoed voor Toerisme*, Nationaal Contact Monumenten, Amsterdam.

Richards, G. (2009) Tourism development trajectories — from culture to creative?, paper presented at the conference *The Asia - Pacific Creativity Forum on Culture and Tourism*, 3-5 June, Jeju Island, Republic of Korea.

Rogerson, C.M. & Collins, K.J. (2015) Beer tourism in South Africa, *Nordic Journal of African Studies*, **24**(3&4), 18-18.

Schnell, S.M. (2011) The local traveler: farming, food, and place in state and provincial tourism guide, 1993-2008, *Journal of Cultural Geography*, **28**(2), 281-309.

Schnell, S.M. (2013) Deliberate identities: Becoming local in America in a global age, *Journal of Cultural Geography*, **30**(1), 55-89.

Shortridge, J.R. (1996) Keeping tabs on Kansas: reflections on regionally based field study, *Journal of Cultural Geography*, **16**(1), 5-16.

Sims, R. (2009) Food, place and authenticity: local food and the sustainable tourism experience, *Journal of Sustainable Tourism*, **17**(3), 321-336.

Slocum, S.L. (2015) Understanding tourism support for a craft beer trail: The case of Loudoun County, Virginia, *Journal of Tourism Planning and Development*, **13**(3), 292-309.

Slocum, S.L., & Curtis, K.R. (2015) Assessing sustainable food behaviours of national park visitors: domestic / on vocation linkages, and their implications for park policies, *Journal of Sustainable Tourism*, **24**(1), 153-167.

Slocum, S.L. & Curtis, K.R. (2018) *Food and Agricultural Tourism: Theory and Best Practice*, London: Routledge.

Slocum, S.L. & Klitsounova, V. (2020) Rethinking tourism in Belarus: The opening of a rural economy, in P. Wiltshier & A. Clarke (eds.), *Community Based Tourism in the Developing World*, London: Routledge, pp. 67-80.

Stone, M. & Nyaupane, G. (2016) Protected areas, tourism and community livelihoods linkages: A comprehensive analysis approach, *Journal of Sustainable Tourism*, **24**, 673-693.

Tepper, S.J. (2002) Creative assets and the changing economy, *Journal of Arts Management, Law, and Society*, **3**(2), 159-168.

Tosun, C. (2000) Limits to community participation in the tourism development process in developing countries, *Tourism Management*, **21**(6), 613-633.

United Nations. (2015) *Transforming our world: The 2030 agenda for sustainable development*, resolution adopted by the General Assembly, 25 September.

United Nations World Tourism Organization. (2018) *Tourism and Culture Synergies*, Madrid, Spain.

Weaver, D.B. (2012) Organic, incremental and induced paths to sustainable mass tourism convergence, *Tourism Management*, **33**, 1030-1037.

Weaver, D.B. & Lawton, L. (2014) *Tourism Management*, Milton, Queensland: John Wiley & Sons.

Wright, W. & Eaton, W.M. (2018) Representing rurality: Cider mills and agritourism, in Slocum, S.L., Kline, C. & Cavaliere, C.T. (eds.), *Craft Beverages and Tourism, Vol.2*, Cham: Palgrave Macmillan, pp. 65-82.

3 Neolocalism, Tourism and Shiplap

Linda J. Ingram

Abstract

As homogenization has led to increased feelings of placelessness and rootlessness, neolocalism offers the opportunity to reconnect with place through nostalgia, rebranding, and marketing of narrative, culture, and heritage through consumption, as well as increasing opportunities for locals. While, conceptually, neolocalism embraces the triple bottom line of sustainability, eclipsing economic interests can give way to a conflicting reality. In this chapter, Magnolia Market at the Silos in Waco, Texas will be highlighted as an example of what can happen when the triple bottom line is sacrificed in favor of economic gain. This renovated complex in a formerly rundown area of Waco has been a financial boon to local tourism related interests while other businesses have faced financial ruin and closure. The aim of this chapter is to expand on established ideas about neolocalism and discuss its relationship to tourism, using the example of the Silos, by proposing and exploring the elements that comprise neolocalism.

Introduction

In this chapter on neolocalism, Magnolia Market at the Silos (the Silos), the brainchild of Waco, Texas, USA entrepreneurs, Chip and Joanna Gaines, will be explored. As hosts of the series *Fixer Upper*, a home renovation series that became an unexpected hit for Home and Garden Television (HGTV), they were inspired to renovate a 2-acre complex in a run-down area of downtown Waco. The complex, which opened in October 2015, houses shops, a bakery, food trucks, and a large plaza complete with picnic tables and benches where families can eat, rest, mingle, and play.

During the first year, over 1.9 million visitors visited Waco to experience the Silos and what it has to offer (Waco History, n.d.). By the second quarter of 2017, hotels in Waco reached 75% occupancy (Country Living, 2016). It is estimated the Silos attracts up to 1.6 million visitors each year with the demand for parking so intense the city added the Silo District Trolley. The popularity of the Silos led to increased attendance at other Waco attractions as well (Burney, 2017). The Gaines are responsible for much of the economic revitalisation Waco is currently experiencing (Halkias, 2017). In short, neolocalism, tourism, and shopping combined to revitalise a city experiencing a lagging economy and a lackadaisical tourist market.

My own history with Waco began in 1967; I lived in the area for 6 years and have returned to visit many times. April of 2018 was my first visit in 12 years, and I found myself wide-eyed at the changes in the downtown area as a result of the Silos. While, as of this writing, there is no available documentation as to specific neolocal principles applied to the development of the complex, many of the principles of neolocalism are evident. However, it was also clear that profit was valued over the triple bottom line of sustainability.

The purpose of this chapter is to incorporate and expand on established ideas about neolocalism and discuss its relationship to tourism. Central to this discussion is the paradox of neolocalism, which while, conceptually, embraces the triple bottom line of sustainability, can, in reality, lead to a conflicting set of results that endow/disenfranchise, empower/disempower and/or liberate/stereotype. The discussion will begin with neolocalism, then segue into brief discussions on place, destination branding and marketing, narratives, the commodification of culture, heritage, symbolic consumption and shopping, and finally, the triple bottom line of sustainability. These themes will then be discussed through the example of the Silos, to be followed by areas for future research and concluding comments.

Understanding the neolocal in tourism

Neolocalism has been embraced as a developing form of tourism as more tourists seek out 'authentic' travel experiences by delving into the 'local' (Mody & Koslowsky, 2019) instead of indulging the tourist gaze when traveling (Wearing et al., 2009). Neolocalism can be understood as a socially constructed idea of place based on a combination of factors, such as historical contexts, nostalgia, and destination marketing, with specific aspects of a locale highlighted and promoted based on the potential for consumption. Burr (2015) argues the primary assumption of social constructionism is that knowledge and understanding are based on human thought, language, and interaction rather than an observable and definable reality. These overlapping concepts provide a pathway to understanding neolocalism as a socially constructed reassertion of reality based on human experience and nostalgia – a desire to recapture a perceived history within a contemporary context – as a way to fend off or repair an increasing sense of rootlessness and loss of sense of place. Or, more simply, the commodification and consumption of 'local' and what it has to offer via the rebranding of place – which is often based on carefully chosen narratives that can omit and/or reframe troublesome and distasteful past events.

As of this writing, research into neolocalism has explored microbreweries (Eberts, 2014; Schnell & Reese, 2003; Slocum et al., 2018), social, environmental, and community resilience (Klink & Denaldi, 2012; Parnwell, 2007; Wilson, 2012), and a variety of other areas (see for example, Brain, 2011; Litzky et al., 2017; Sipe, 2019). Exploring the 'local' in destinations via neolocal tourism remains largely unexamined (Mody & Koslowsky, 2019). While the 'deliberate seeking out of regional lore and local attachment' (Shortridge, 1996: 10) in the quest for 'authentic' experiences has become popular with tourists seeking to avoid the pitfalls of traditional mass tourism, the reality of neolocalism is much more complex. While participation in neolocalism can be viewed as a form of resistance to homogenisation and rootlessness, any notion that neolocal tourism is an answer to mass tourism can and should be met with scepticism (Mody & Koslowsky, 2019).

While neolocal tourism development is increasing, it is not always the result of strategic policy and planning. Mody and Koslowsky (2019) contend that many destinations are unprepared for the introduction of neolocal tourism, which can result in a variety of issues including loss of place identity for locals, overcrowding, and increased security concerns. Loss of place identity and community are often the result of travellers seeking out 'authentic' experiences (Mody & Koslowsky, 2019), which can also create long-term housing shortages when local residences are turned into short-term tourist accommodations (Longhi, 2019). Haphazard and unplanned development of this nature can lead to the negative impacts associated with mass tourism (Mody & Koslowsky, 2019).

Place and destination branding

Celebration of place is at the heart of neolocalism. Places endowed with attachment and meaning are associated with sense of place (Brehm et al., 2013). Tuan (1977) argues sense of place is a social construction, dynamic in nature, that can refine human feeling and perception. Hede and Watne (2013) suggest sense of place depicts the relationship between place, attachment, and consumers. The ongoing deterioration of sense of place has contributed to feelings of loss of community, tradition, family cohesiveness, and stability (Arefi, 2007). As technology continues to create 'placeless communities' (Schnell, 2013: 55) through social media, gaming, and other online platforms, in addition to the homogenising effects of globalisation, reconnecting to 'geographical identities' (Shortridge, 1996: 10) has become increasingly important as we seek to consciously reaffirm and re-establish sense of place. Holtkamp et al. (2016) refer to neolocalism 'as a conscious effort by businesses to foster a sense of place based on attributes of the community' and 'an active, conscious attempt to create a new sense of place' (p. 66). This combination of factors provides marketers with a foundation for brand strategies through the use of narratives associated with sense of place. Therefore, celebration of place can be understood as the promotion of place through narrative.

Promotion of place can be found under the umbrella of destination branding. While the definition of destination branding is a contested one (Anholt, 2010), for the purposes of this chapter, the concept will be understood simply as 'a way to communicate a destination's unique identity by differentiating a destination from its competitors' (Morrison & Anderson cited in Qu et al., 2011: 466). The complexity, of course, lies in the differentiation. Branding is about identity – a destination's personality – which is fundamental to communicating the distinctiveness of that brand, as well as determining the target market, which is key to motivating consumers to identify with the brand and integral to its success and popularity (Qu et al., 2011). Research indicates consumers also have substantial influence over brand creation. Some have argued consumers contribute through brand interaction, which, in turn, can affect development of their own identities and lifestyles (Black & Veloutsou, 2017). The use of brands for the symbolic construction of self-identity is well documented (Elliott & Wattanasuwan, 1998). Social identity theory postulates that consumers even appropriate aspects of their preferred self-concept from specific social groups to which they belong/want to belong through the brands they use (Tajfel & Turner, 1986).

If branding is about who you are, then marketing is about communicating what you are. Traditionally, tourism destinations have been marketed as products and services to be consumed, with quantifiable assets used to enhance profit. This perspective encourages an objective view of place as a product, which fails to take into account place meaning and contested relationships, but rather encourages the idea that destinations should focus their resources on tourist satisfaction instead of attending to the needs of the community (Lichrou et al., 2008: 29). The negative impacts of this long-time strategy have been well documented (Singh, 2017). Lichrou et al. (2008) argue that the marketing of tourism destinations should instead be framed through the lens of distinctive narratives, which promotes the idea that the needs of local stakeholders, preserving natural resources, and maintaining cultural distinctiveness must all be taken into consideration when developing and applying a destination's brand. This course of action provides a pathway to establishing tourism development that can serve the needs of both local stakeholders, tourism providers and tourists by maintaining and enhancing the elements connecting human experience to Shortridge's 'geographical identities' (1996: 10) while simultaneously protecting and sustaining the destination's environment. Neolocalism is intertwined with marketing and branding – self, group, and destination – as both function by establishing an emotional connection via socially constructed narratives designed to fulfil consumer desire.

Narratives

Somers (1997) suggests we gain insight into the social world and our place in it, then construct our own social identities through the use of narratives. No one narrative is universally accepted, nor does everyone accept or believe in the same group(s) of narratives. Ontological narratives (Somers, 1997) are 'the stories that social actors use to make sense of…their lives…locating ourselves in narratives endows us with identities, however multiple, ambiguous, ephemeral or conflicting they may be' (p. 84). These identities are enhanced through self and social symbolism (Elliott & Wattanusuwan, 1998). Individual and social identities are linked through cultural symbols with which they feel a connection (Black & Veloutsou, 2017). One of the more prevalent narratives in contemporary culture, and a key aspect of this discussion, began in the 1960s with the idea of rootlessness due to the effects of globalisation (Flack, 1977; Mayhar, 1999). Mody and Koslowsky (2019) posit this sense of rootlessness has led to the increase in travellers seeking 'authenticity' through neolocal travel, while Poria et al. (2003) link the search for 'authenticity' and rootedness to place. Unfortunately, the quest for reclaiming rootedness can result in the commodification of the place sought for 'authenticity' (Williams & Shaw, 1992).

Cultural tourism commodified

Marx views a commodity as 'an object outside us, a thing that by its properties satisfies human desires' (Baillie et al., 2010: 53). Market capitalism takes this idea further by promoting continuous acquisition and consumption as a way to locate the self in contemporary society, in addition to constructing and maintaining identities while participating in the process of hegemony as people differentiate themselves from others (Parker, 2011).

A universally accepted definition of culture has remained elusive despite decades of debate and multiple analytical perspectives. Current understandings of culture appear to be dependent upon the field in which it is being studied, with actively contested viewpoints even within disciplines. Shifting understandings of, and uses for, culture have provided a variety of concepts and applications for research and interpretation, which are valuable in furthering debate and study highlighting the complex nature of culture but doing little to move toward consensus (Baldwin et al., 2006). The idea of culture has evolved from perceptions of it as static and unchanging to understanding its usefulness as a method for studying the intersections of power, inequality, and history (Baldwin et al., 2006). This chapter will borrow from Linton (1945) who describes culture as the learned and shared behaviours passed from generation to generation.

'Tourism is seen to be a major part of how people, both tourists and locals alike, perceive, experience, and create places' (Lew, 2016: 172). The link between tourism and place is strong, yet seldom explored in spite of the link between place and culture (Smith, 2015). Cultural tourism has been equally difficult to define. Smith (2015) describes cultural tourism as an activity associated with consumption of heritage, education, the arts, agriculture, gastronomy, history, and heritage. In other words, 'culture, like God and politics, is everywhere' (Adair, 1992 as cited by Smith, 2015: 222). The same might be said of tourism. International tourism arrivals reached a record 1.19 billion in 2015 with a total of 18 billion expected by 2030 (Statista, 2018: para. 2). Shepherd (2002) argues the rise of tourism inevitably leads to cultural commodification as 'tourist purchases are fueled by a desire to possess a mark, rather than any genuine interest in local cultural traditions or beliefs' (p. 185).

Symbolic consumption and shopping

Consumption is a symbolic act (Wattanasuwan, 2010). Symbolic consumption is socially constructed, with no external basis for evaluation. Research indicates it is malleable, subject to the idiosyncrasies of marketers, advertisers, consumers, and societal changes. The transitory nature of symbolic consumption leaves it open to interpretation depending on who is doing the consuming and their personal symbols. Hoechsmann (2010) argues, 'consumption has become an important determinant of our social selves, a primary site of self- and group-identity' (p. 23).

Purchasing goods is a method of identity creation, representation, belonging, and status: we are what we own (see for example: Ekinci et al., 2013; Wattanasuwan, 2010) – a phenomenon echoed in social identity theory, as noted above. Symbolic consumption is also one of the methods used to create connections with others (Wattanasuwan, 2010). 'Consumption is not simply an act of compliance but is rather a complex mediation of self- and group-identity with everyday needs and desires' (Hoechsmann, 2010: 24). The bottom line here is that the self is socially constructed in a manner designed to contribute to one's well-being, which, typically, involves substantial symbolic consumption. If neolocalism is recognised as a response to a postmodern era fraught with fragmentation and rootlessness, participation in the neolocal can be understood as a response to the fracturing of the self as consumers seek to reconstruct and solidify notions of that self through symbolic consumption.

Shopping and tourism have become almost inexorably linked to the point that some destinations are developing additional shopping facilities for tourists as a way to increase expenditures while increasing employment opportunities for locals. Tourists often spend more money on shopping than other aspects of travel, such as lodging, food, or entertainment. Research indicates tourists tend to buy items from well-known brands unique to the destinations they cannot access at home (Wong & Wan, 2013). Lin and Mao (2015) identify the importance of souvenir shopping for personal use or gifting to family and friends as 'material objects identifying place' (p. 20).

The connection between symbolic consumption and tourism has been well documented and is understood as another way of defining and enhancing both identity and self-image (Ekinci et al., 2013). However, this quest for identity and self-image enhancement has also heightened the negative impacts associated with mass tourism (Namberger et al., 2019). Hjalager (2000) notes the ethical concerns associated with overconsumption and overuse of resources in mass tourism, arguing for sustainable tourism development. However, the path to sustainable tourism development is problematic at best (Girard & Nijkamp, 2009).

Neolocalism and sustainability

The triple bottom line of sustainability includes environmental, economic, and socio-cultural factors (Holtkamp et al., 2016), all of which correlate with neolocalism. Sustainable tourism development, yet another contested topic (García et al., 2018), has been promoted as a way to mitigate the negative impacts of mass tourism and overtourism by utilising the triple bottom line (Wu et al., 2019). While the idea of sustainable tourism development is conceptually desirable, success hinges on proper implementation, which, in some instances, has been fragmented and ineffective, often due to ineffective governance (Farmaki, 2015). In other cases, destinations have been slow to implement sustainable tourism development due to lackadaisical efforts as profit overshadows the triple bottom line (Dodds & Butler, 2009). There is also an argument to be made that sustainable tourism development is an oxymoron (Wall, 2016).

Holtkamp et al. (2016), citing Curtis (2003), argue that sustainability is fundamental to neolocalism as 'the road to environmental sustainability lies in the creation of local, self-reliant, community economies' (p. 83). Local entrepreneurs have been recognised as significant sources of economic growth (Chatman et al., 2008; Henderson, 2002) and are foundational to neolocalism. Graefe et al. (2018) posit neolocalism can also promote local sourcing and influence environmentally-friendly production techniques and consumption patterns, as well as encourage support for local causes and charities. Corporate social responsibility (CSR) is typically associated with large corporations. However, many small businesses have adopted the tenets of CSR, which correlate to the values of the owner(s) and community while gaining a competitive advantage as consumers increasingly seek out businesses perceived to embrace local values (Holtkamp et al., 2016), another aspect of neolocalism.

Neolocal tourism's focus on sustainability includes the importance of local production and artisans, sense of place, community engagement, and equity. The development and implementation of neolocal tourism is both complex and challenging. The example of the Silos is a common one in neolocalism. The remainder of the discussion highlights what happens when some aspects are implemented while others are left out of the equation.

Neolocalism and the Silos

Texas is one of many places fighting to retain its cultural image and sense of place, which makes it an ideal location for a case study in neolocal tourism. Texas identity looms large, and many see their way of life as under attack and in danger of disappearing unless they take action (Fernandez, 2016). The prolific historian, T.R. Fehrenbach, was one of the Texans who first led the charge. Although his classic *Lone Star* (1968) was meant to simply portray historical events as they happened, in the five decades since its publication, the book has served to reignite a passion for Texas mythology, identity, and history (Martin, 2013). For Fernandez (2016), 'the line between the myth of Texas and the reality of Texas is razor thin' (para. 6). Texas' natural and built environments, charm, hospitality, cuisine, football, and patriotism are central to its cultural composition, narrative, and sense of place. Texas history is also fundamental to its distinctiveness.

In 1993, a standoff between federal agents and the Branch Davidians, 13 miles east-northeast of Waco, led to the deaths of 76 people. While Waco had never been a mecca for tourism, the city somehow became connected with the siege in people's minds, resulting in both new businesses and tourists staying away. The election of George W. Bush in 2000, whose ranch is located 25 miles away in Crawford, led to a slight increase in tourism for those wanting to see the 'Western White House' (Waco History, n.d.). However, it was the addition of *Fixer Upper* to the HGTV line-up in 2013 that turned Waco tourism into a force with which to be reckoned. The popularity of *Fixer Upper* influenced the development of the Magnolia brand used throughout the Gaines' entrepreneurial empire. The brand can be found in many retail outlets, as well as online, but its home is the Silos, where place, consumption, and tourism intersect with neolocalism.

The narrative of the Silos begins with the Brazos Valley Cotton Oil Company who had the structures built in 1950. Even though ownership changed hands over time, the two 120-foot tall silos were in use through the 1980s. They are synonymous with the city's agricultural past and are among Waco's most well-known landmarks (Sawyer, n.d.). Purchased by the Gaines in 2014 (Sawyer, n.d.), the Silos narrative is a complex one that weaves beliefs about gender roles, marriage, family, tradition, Christianity, and the importance and power of home and hearth with time-honoured Texas narratives embracing notions of independence, self-reliance, resilience, strength, individuality, community, and pride – the same components of the Magnolia brand – with Joanna's 'farmhouse chic' or 'rustic chic' style. The Silos, HGTV'S *Fixer Upper*, and other businesses in the Gaines entrepreneurial empire all embrace this same narrative and branding.

The branding of the Silos is a reflection of its distinctiveness, as well as the personalities of the owners and their entrepreneurial ventures. The strength of the brand is evident in a variety of ways, including the number of visitors and the desire of many to self-identify with the Gaines, as well as the financial power of their expanding empire. The brand's marketing success is undeniable: by 2017, the Silos averaged approximately 30,000 visitors a week (Loudenback, 2018). While the Gaines have a strong marketing and advertising campaign, the Silos, as well as their other business ventures, are a popular topic for many publications and websites providing substantial free advertising and brand promotion. Chip and Joanna's narrative of a hard-working young couple devoted to family who are both in love and in like with each other provides a form of brand equity that is one of the driving forces behind the show's ratings, the popularity of the Silos, and the ever-expanding Gaines entrepreneurial holdings. The couple are relatable to their viewers and receive an added social media boost from fans who watch the Gaines children grow and look for Joanna's next endeavour while seeing 'what's new' at the couple's farm. This perceived access enhances the idea fans 'know the hosts', thereby strengthening and solidifying the connection (Country Living, 2016), resulting in a shared narrative (brand).

When I visited the Silos in April 2018, it was apparent the types of goods and food sold are virtually identical to those found throughout the home goods retail world. The fulfilment of consumer desire seems to revolve around purchasing items associated with the brand, which supports the assertion that brand consumption is based upon the usefulness of that brand to consumers in fulfilling material, emotional, and psychological needs (Ekinci et al., 2013). One of the days I visited was an unusual one for the area – sleet, wind, and freezing cold – so I assumed tourists would likely stay home. I was wrong. For example, visitors were in a line that started at the front door of the Silos Baking Co., wrapped around the corner, and then snaked down the block (just to buy cupcakes!), the vast majority without umbrellas or weather appropriate clothing. Exhaust belching tour buses dropped off a seemingly endless number of tourists who scattered throughout the Silos upon disembarking. The power of identity creation and enhancement, belonging and status, combined with symbolic consumption to mediate the self in daily life, cannot be overstated in this instance. While people-watching on that rainy day, I observed many visitors gushing over how a certain product was so 'me!' or the pride already in evidence as a shopper described how a specific item would enhance their home and the overflowing excitement about how others would now perceive that person who had not only visited the Silos, but, after purchasing household goods, would have stories to relate to friends and family, thereby enhancing their status.

Shopping and watching *Fixer Upper* are not the only ways aficionados connect with the brand. Membership in this social world represents 'a unique scheme of life in which members share in a special set of meanings and in which various cultural elements…are created and made meaningful by social world members and serve to set the social world apart from other social worlds' (Scott & Godbey cited in Kyle & Chick, 2007: 213). For example, one of the events the Silos hosts throughout the year is the Annual Magnolia Silobration in October described as 'three days of pure joy' (Mead, 2018, para. 2). Travellers from all over the USA descend on the Silos to enjoy a weekend of music, food, and over 80 vendors, as well as an appearance by Chip and Joanna. For some, no effort is too extreme: two friends drove 18 hours for the experience. Another participant chirped, 'Chip and Joanna Gaines created this dream for all of us to get to know each other, and you feel like everybody is family here so we wanted that. And that's why we flew all the way from Sun Valley, California to experience Silobration' (Trotter, 2018, para. 5). Membership in this social world is coveted among aficionados: 50,000 attended Silobration in 2018 (Trotter, 2018). The 2019 attendance numbers were not available at this writing, however hotels in the areas were once again sold out weeks before Silobration, and at least two dozen non-Silobration events were scheduled by businesses in the downtown area seeking to ride on the coattails of the popular festival (Smith, 2019).

Not all visitors are as upbeat. Some who were not pleased with their Silos experience have voiced their opinions on TripAdvisor calling it an 'overpriced Hobby Lobby', 'did not see the fascination' (TripAdvisor, n.d., i) or labelled it 'disappointing' and 'big hype for nothing' (TripAdvisor, n.d., ii). Some residents decry the changes that have occurred in, and to, Waco as a result of the uptick in tourism since the opening of the Silos, citing higher property taxes, increased traffic and congestion, ongoing construction (both in the area of the Silos and the number of households remodelling to get Joanna's signature look), loss of sense of place, insufficient police presence (some residents say the city is not keeping up with current needs for protection), and increasing housing costs (Saegert, 2019). The popularity and influence of the *Fixer Upper*/Silos narrative extends beyond home renovation and retail: Waco's Baylor University has developed 'Shiplap' (named after Joanna's love and use of shiplap in renovations) to assist accounting students in developing their career path (Burney, 2017). Another phenomenon, known as 'The Gaines Effect' or the 'Fixer Upper Effect', is the employment of around 600 people by the Gaines holdings in Waco (Kapusta, 2018). Given its vast retail presence, influence and popularity, there is an argument to be made that the Gaines brand has evolved into a brand community (Stratton & Northcote, 2014).

The power and influence of this brand at the Silos complex is staggering. There are a variety of factors pointing to mass tourism – attendance numbers, ongoing expansion of retail space, and on-site special events – and all are indicative of large-scale commodification and consumption. The brand's online presence is extensive, as well, and goes far beyond the Magnolia website. The Silos, the online presence, a magazine, household goods in multiple retail outlets, and a new TV network reflect the intensity of the narrative's resonance with consumers. The Silos also celebrates Texas history and heritage, family, friends, community, Christianity, and the Gaines narrative. The economic and cultural ripple effect of the Silos on the local economy is undeniable; its influence has spread beyond two blocks in downtown Waco. The Silos narrative could be interpreted as a grassroots campaign to enhance and expand a tightly focused narrative promoting a specific set of behaviours and values. Membership in this social world demonstrates consumers' desire to seek out community while simultaneously seeming to covet the individuality of the combined Gaines and Texas narratives.

All that glitters isn't gold. During my April 2018 visit, there was little evidence of concern for negative environmental impacts. There were no recycling bins. Food trucks with single use to-go containers and the accompanying trash were an issue, as well as the energy it takes to transport and run them. Mass consumption and commodification are the antithesis of sustainable tourism development. The conversion of the Silos into a retail endeavour is a success, but it came at a price for its neighbours; some of the businesses surrounding the complex are not faring well. Parking and overcrowding have led to the sale of other businesses in the area, decreasing economic diversity. Evidence of local artisans' work was also elusive as the retail products sold at the Silos are mass produced. By disenfranchising neighbouring business owners, minimal utilisation of local artisans, a lack of environmental controls or sustainable practices, and borderline overtourism, the Silos is indicative of what can happen to neolocalism when the financial bottom line receives greater focus than the triple bottom line.

Symbolic consumption at the Silos shows no sign of abating. In June 2019, the Gaines announced a $10.4 million expansion of the Silos that will include additional shops, landscaping, gardens, and brick pathways along with a whiffle ball field and splash pad for children. Buildings outside of the current complex will be incorporated into the Silos as additional retail space. The Second Presbyterian Church will be relocated to the site and used for seminars and additional seating (Saegert, 2019). Expansion of the Silos is not the only way the brand is growing. In November 2018, Chip announced the couple is developing their own lifestyle-oriented TV network with Discovery, Inc. (Strohm, 2018). The couple recently purchased an old building in downtown Waco and announced plans to develop it into a boutique hotel scheduled to open in 2021. They are also the owners of Cottonwood Castle: plans for it are forthcoming (Aalagha, 2019). The Magnolia Press Coffee Co. opened October 17, 2019 (Netkin, 2019). While the success of the Silos and associated businesses have been a welcome financial windfall for Waco and the Gaines, the lack of attention to sustainable tourism development is a troubling example of large profits, once again, overshadowing socio-cultural and environmental concerns. The notoriety the brand has achieved offered the opportunity to fully embrace neolocalism and showcase the

Silos as an example of sustainable tourism development. Instead, financial gain won the day.

Conclusion

The objective of this chapter was to initiate a discussion about neolocalism that incorporated and expanded on current notions about the phenomenon and link it to tourism through the example of the Silos in Waco, Texas, USA. Place, as a socially constructed phenomenon, is foundational to neolocalism. In an era where feelings of placelessness and rootlessness are increasingly common, nostalgia and historical events provide a framework for neolocalism through the rebranding and marketing of narrative, culture, and heritage, as well as symbolic consumption and shopping. The discussion has shown that neolocalism is a complex phenomenon with interconnected components grounded in a multi-disciplinary array of theories and concepts.

There are many areas open to future research; only a few will be suggested here. This discussion has merely opened the door to understanding the complexity and diversity of neolocal tourism and through the lens of a single destination. The first step would be to determine the influence and effects of neolocalism on tourism and look for similarities, differences, and patterns. Another avenue of research would be to explore how neolocal tourism affects, and is perceived by, local stakeholders. As in the example of the Silos, it is apparent that while the retail complex has been successful for the Gaines and the city of Waco as a whole, it has been detrimental to some of the business owners in the area, as well as stakeholders who face additional economic challenges as a result of rising home prices, rents, property taxes, and overcrowding, along with increasing traffic and parking challenges.

Neolocalism is used by stakeholders as a way to enhance and preserve the local through place, branding, narrative, and sustainable commodification for consumption by the individuals who seek to enhance self-image, status, belonging, and community. While the concept of neolocalism carries the potential to protect and preserve the local, there are also pitfalls if the desire for financial gain overshadows the protection of established cultural, natural, and built environments. Hopefully, this chapter has opened the door to understanding neolocalism not as a stand-alone phenomenon, but instead as a complex series of interconnected components.

References

Aalagha, H. (2019) 'We love Waco:' Chip and Joanna Gaines reveal Magnificent plans for the city's future, *KCEN TV*, 12 November. Available at https://www.kcentv.com/article/news/local/mclennan-county/chip-and-joanna-gaines-talk-mag-nificent-plans-for-the-citys-future/500-cb8d8bfc-2fed-475e-9c62-604a01bab5f8. (Accessed: 12/11/2019).

Anholt, S. (2010) Definitions of place branding – Working towards a resolution, *Journal of Place Branding and Public Diplomacy*, **6**(1), 1-10.

Arefi, M. (2007) Non-place and placelessness as narratives of loss: Rethinking the notion of Place, *Journal of Urban Design*, **4**(2), 179-193.

Baillie, B., Chatzoglou, A. & Taha, S. (2010) Packaging the past: the commodification of Heritage, *Heritage Management*, **3**(1), 51-71.

Baldwin, J.R., Faulkner, S.L. & Hecht, M.L. (2006) A moving target: The illusive definition of culture, in J.R. Baldwin, S.L. Faulkner, M.L. Hecht & S.L. Lindsley (eds.), *Redefining Culture*, New York, NY: Routledge, pp. 27-50.

Black, I. & Veloutsou, C. (2017) Working consumers: Co-creation of brand identity, consumer identity and brand consumer identity, *Journal of Business Research*, **70**, 416-429.

Brain, T.J. (2011) *Examining the Portland music scene through neo-localism*, Master's thesis, Portland State University, Portland.

Brehm. J.M., Eisenhauer, B.W. & Stedman, R.C. (2014) Environmental concern: Examining the role of place meaning and place attachment, *Society and Natural Resources*, **26**(5), 522-538.

Burney, L. (2017) Accounting student transformations: Baylor University takes a few cues from HGTV's Fixer Upper to help students find the right accounting career path, *Strategic Finance*, **99**(2), 36-42.

Burr, V. (2015) *Social Constructionism*, 3rd edn, London: Routledge.

Chatman, D., Altman, I. & Johnson, T. (2008) Community entrepreneurial climate: An analysis of small business owners' perspectives in 12 small towns in Missouri, USA, *Journal of Rural and Community Development*, **3**, 60-77.

Country Living (2016), This is why you're so obsessed with 'Fixer Upper' *Country Living*, 8 July. Available at www.countryliving.com/life/entertainment/news/a39131/why-you-are-obsessed-with-fixer-upper/. (Accessed: 18/11/2018).

Dodds, R. & Butler, R. (2009) Barriers to implementing sustainable tourism policy in mass tourism destinations, *Tourismos*, **5**(1).

Eberts, D. (2014) Neolocalism and the branding and marketing of place by Canadian Microbreweries, in M. Patterson & N. Hoalst-Pullen (eds.), *The Geography of Beer: Regions, Environment, and Societies*, Dordrecht: Springer, pp. 189-199.

Ekinci, Y., Sirakaya-Turk, E. & Preciado, S. (2013) Symbolic consumption of tourism destination brands, *Journal of Business Research*, **66**(6), 711-718.

Elliott, R. & Wattanusuwan, K. (1998) Brands as symbolic resources for the construction of identity, *International Journal of Advertising,* **17**(2), 131-144.

Farmaki, A. (2015) Regional network governance and sustainable tourism, *Tourism Geographies,* **17**(3), 385-407.

Fehrenbach, T.R. (1968) *Lone Star: A History of Texas and the Texans,* 1st edn, Ontario, Canada: Collier Books.

Fernandez, M. (2016) What Makes Texas Texas, *New York Times,* 8 May. Available at https://www.nytimes.com/2016/05/08/us/what-makes-texas-texas.html. (Accessed: 10/05/2019).

Flack, W. (1997) American microbreweries and neolocalism: "Ale-ing" for a sense of place, *Journal of Cultural Geography,* **16**(2), 37-53.

García, M.O., Amorim, E. & Korstanje, M. (2018) The power of knowledge in the struggle for sustainable tourism, in H. Séraphin & E. Nolan (eds.), *Green Events and Green Tourism: An International Guide to Good Practice,* London: Routledge.

Girard, L.F. & Nijkamp, P. (2016) Narrow escapes: pathways to sustainable Local cultural Tourism, in L.F. Girard & P. Nijkamp (eds.), *Cultural Tourism and Sustainable Local Development,* London: Routledge, pp. 19-28.

Graefe, D., Mowen, A. & Graefe, A. (2018) Craft beer enthusiasts' support for neolocalism and environmental causes, in S.L. Slocum, C. Kline & C.T. Cavaliere (eds.), *Craft Beverages and Tourism, Vol. 2,* Cham: Palgrave MacMillan, pp. 27-47.

Halkias, M. (2017) Waco's Magnolia Market at the Silos is now more popular with tourists than the Alamo, *Dallas Morning News,* 20 November. Available at https://www.dallasnews.com/business/retail/2017/11/20/wacos-magnolia-market-silos-now-popular-tourists-alamo. (Accessed: 4/04/2019).

Hede, A-M. & Watne, T. (2013) Leveraging the human side of the brand using a sense of place: Case studies of craft breweries, *Journal of Marketing Management,* **29**(1-2), 207-224.

Henderson, J. (2002) Building the rural economy with high growth entrepreneurs, *Economic Review,* Federal Reserve Bank of Kansas City, **Q III**, 45-70.

Hjalager, A.M. (2000) Consumerism and sustainable tourism, *Journal of Travel & Tourism Marketing,* **8**(3), 1-20. Hoechsmann, M. (2010) Rootlessness, Reenchantment, and Educating Desire, in J.A. Sandlin & P. McLaren (eds.), *Critical Pedagogies of Consumption: Living and learning in the shadow of the 'Shopocalypse',* New York, NY: Routledge, pp. 23-35.

Holtkamp, C., Shelton, T., Daly, G., Hiner, C.C. & Hagelman III, R.R. (2016) Assessing neolocalism in microbreweries, *Papers in Applied Geography,* **2**(1), 66-78.

Kapusta, M. (2018) How much Chip and Joanna have boosted Waco's economy and why some aren't happy about it, *Showbiz Cheat Sheet,* 28 October. Available at https://www.cheatsheet.com/entertainment/how-much-have-chip-and-joanna-gaines-boosted-wacos-economy.html/. (Accessed: 17/11/2018).

Klink, J. & Denaldi, R. (2012) Metropolitan fragmentation and neo-localism in the periphery: revisiting the case of Curitiba, *Urban Studies,* **49**(3), 543-561.

Kyle, G. & Chick, G. (2007) The social construction of a sense of place, *Leisure Sciences,* **29**(3), 200-225.

Lew, A. (2014) Introduction to special issue – Cultural geographies of tourism: image, identity and place, *Tourism Geographies,* **16**, 171-173.

Lichrou, M., O'Malley, L. & Patterson, M. (2008) Place-product or place narrative(s)? Perspectives in the marketing of tourism destinations, *Journal of Strategic Marketing,* **16**(1), 27-39.

Lin, L. & Mao, P.C. (2015) Food for memories and culture – A content analysis study of food specialties and souvenirs, *Journal of Hospitality and Tourism Management,* **22**,19-29.

Linton, R. (1945) *The Cultural Background of Personality,* New York, NY: D. Appleton-Century Co.

Litzky, B.E., Andersson, L. & Smith, W.P. (2017) Local entrepreneurs: conduits of neo-localism and sustainable urban livelihoods, in *Academy of Management Proceedings* **2017**(1), 10968.

Longhi, L. (2019) Sedona's housing woes threaten its sense of community. Some blame vacation rentals, *The Arizona Republic,* 29 January. Available at https://www.azcentral.com/story/news/local/arizona/2019/01/29/sedona-faces-housing-shortage-short-term-vaction-rentals-take-over/2342431002/. (Accessed: 15/05/2019).

Loudenback, T. (2018) Millions of tourists are flocking to Waco, Texas, to see 'Fixer Upper' stars Chip and Joanna Gaines small home town – here are all the best things to do there, *Business Insider,* 3 April. Available at https://www.businessinsider.com/hgtv-chip-and-joanna-gaines-made-waco-texas-top-travel-destination-2018-1. (Accessed: 18/11/2018).

Martin, D. (2013) T.R. Fehrenbach, Historian, Dies at 88: Chronicler of Larger-Than-Life Texas, *New York Times,* 4 December. Available at https://www.nytimes.com/2013/12/04/us/tr-fehrenbach-historian-of-a-larger-than-life-texas-dies-at-88.html. (Accessed: 10/05/2019).

Mayhar, A. (1999) Non-place and placelessness as narratives of loss: Rethinking the notion of Place, *Journal of Urban Design,* **4**(2), 179-193.

Mead, T. (2018) 6 Things you should know about Chip and Joanna Gaines's Magnolia Silobration, *House Beautiful,* 17 October. Available at https://www.housebeautiful.com/lifestyle/a23878016/chip-joanna-gaines-magnolia-silobration/. (Accessed: 30/11/2018).

Mody, M. & Koslowsky, K. (2019) Panacea or peril? The implications of Neolocalism as a more intrusive form of tourism, *Boston Hospitality Review,* **7**(1).

Namberger, P., Jackisch, S., Schmude, J. & Karl, M. (2019) Overcrowding, overtourism and local level disturbance: how much can Munich handle?, *Tourism Planning & Development,* **16** (4), 452-472.

Netkin, S. (2019) Joanna Gaines just opened her Magnolia Press Coffee Shop in Texas, 17 October. Available at https://www.delish.com/food-news/a28306789/joanna-gaines-announces-coffee-shop-magnolia-press/. (Accessed: 1/11/2019).

Parker, H.N. (2011) Toward a definition of popular culture, *History and Theory,* **50**, 147-170.

Parnwell, M.J. (2007) Neolocalism and renascent social capital in northeast Thailand, *Environment and Planning D: society and space,* **25**(6), 990-1014.

Poria, Y., Butler, R. & Airey, D. (2003) The core of heritage tourism, *Annals of Tourism Research,* **30**(1), 238-254.

Qu, H., Kim, L.H. & Im, H.H. (2011) A model of destination branding: Integrating the concepts of the branding and destination image, *Tourism Management,* **32**(3), 465-476.

Saegert, R. (2019) $10.4 million Magnolia Silos expansion to feature recreation, shopping, relocated church, *Waco Tribune-Herald,* 19 June. Available at https://www.wacotrib.com/news/business/million-magnolia-silos-expansion-to-feature-recreation-shopping-relocated-church/article_8204fa20-2cb5-53c7-874d-5345a76fbeb0.html. (Accessed 30/06/2019).

Sawyer, A. (n.d.) Brazos Valley Cotton Oil Mill, *Waco History*. Available at https://wacohistory.org/items/show/97. (Accessed: 18/11/2018).

Schnell, S.M. (2013) Deliberate identities: Becoming local in America in a global age, *Journal of Cultural Geography,* **30**(1), 55-89.

Schnell, S.M. & Reese, J.F. (2003) Microbreweries as tools of local identities, *Journal of Cultural Geography,* **21**(1), 45-69.

Shepherd, R. (2002) Commodification, culture and tourism, *Tourist Studies,* **2**(2), 183-201.

Shortridge, J.R. (1996) Keeping tabs on Kansas: reflections on regionally based field study, *Journal of Cultural Geography,* **16**(1), 5-16.

Singh, T.V. (2017) Tourism and development: not an easy alliance, in R.N. Ghosh & M.A.B. Siddique (eds.), *Tourism and Economic Development: Case Studies from the Indian Ocean,* London: Routledge, pp. 30-41.

Sipe, L.J. (2019) Leveraging neo-localism for experience innovation: A case study of an urban park and entertainment venue, *Journal of Themed Experience and Attractions Studies,* **1**(1), 29-37.

Slocum, S.L., Kline, C. & Cavaliere, C.T. (2018) *Craft Beverages and Tourism, Volume 2,* Cham: Palgrave Macmillan.

Smith J.B. (2019) Silobration organizers eye traffic, parking for up to 40,000 visitors, *Waco Tribune-Herald,* 16 October. Available at https://www.wacotrib.com/business/silobration-organizers-eye-traffic-parking-for-up-to-visitors/article_906507e6-8a81-5a16-b40b-fa604928875d.html. (Accessed: 12/11/2019).

Smith, S. (2015) A sense of place: Place, culture and tourism, *Tourism Recreation Research,* **40** (2), 220-233.

Somers, M.R. (1997) Deconstructing and reconstructing class formation theory: Narrativity, relational analysis, and social theory, in J.R. Hall (ed.) *Reworking Class*, Ithaca, NY: Cornell University Press, pp. 73-106.

Statista. (2018) Global travel and tourism industry – Statistics & Facts. *statista*, 21 August. Available at https://www.statista.com/topics/962/global-tourism/. (Accessed: 25/10/2019).

Stratton, G. & Northcote, J. (2014) When totems beget clans: The brand symbol as the defining marker of brand communities, *Journal of Consumer Culture*, **16**(2), 493-509.

Strohm, E. (2018) Chip and Joanna Gaines are returning to TV with their own Discovery network, *People*, 10 November. Available at https://people.com/home/chip-joanna-gaines-tv-network-return-to-television/. (Accessed: 17/11/2018).

Tajfel, J.C. & Turner, H. (1986) The social identity theory of intergroup behavior, *Psychology of intergroup relations*, **5**, 7-24.

Trip Advisor. (n.d., i) Silobration (at Magnolia Farms/Silos)…Is it worth going? Available at https://www.tripadvisor.com/ShowTopic-g56833-i1450-k10618483-Silobration_at_Magnolia_Farms_Silos_Is_it_worth_going-Waco_Texas.html. (Accessed: 17/11/2018).

Trip Advisor. (n.d., ii). Magnolia Market. Available at https://www.tripadvisor.com/Attraction_Review-g56833-d8733906-Reviews-Magnolia_Market-Waco_Texas.html. (Accessed: 30/11/2018).

Trotter, M. (2018) 4th annual Magnolia Silobration brings 50,000 people to Waco this weekend, *Fox 44 News*, 18 October. Available at https://www.fox44news.com/news/local-news/local/4th-annual-magnolia-silobration-brings-50000-people-to-waco-this-weekend/. (Accessed: 17/11/2018).

Tuan, Y-F. (1977) *Space and Place – The perspective of experience*, Minneapolis, MN: University of Minnesota Press.

Waco History. (n.d.) Available at https://wacoheartoftexas.com/wp-content/uploads/2017/05/Waco-History-2017.pdf. (Accessed: 18/11/2018).

Wall, G. (2016) Tourism and development: Towards sustainable outcomes, in L.F. Girard & P. Nijkamp (eds.), *Cultural tourism and sustainable local development*, London: Routledge, pp. 49-64.

Wattanasuwan, K. (2010) The self and symbolic consumption, *Journal of American Academy of Business*, **6**(1), 179-184.

Wearing, S., Stevenson, D. & Young, T. (2009) *Tourist Cultures: Identity, place and the traveller*. Newbury Park, CA: SAGE.

Williams, A.M. & Shaw, G. (1992) Tourism research: a perspective. *American Behavioral Scientist*, **36**(2), 133-143.

Wilson, G. (2012) *Community Resilience and Environmental Transitions*, London: Routledge.

Wong, I.A. & Wan, Y.K.P. (2013) A systematic approach to scale development in tourist shopping satisfaction: Linking destination attributes and shopping experience, *Journal of Travel Research*, **52**(1), 29-41.

Wu, K.J., Zhu, Y., Chen, Q. & Tseng, M.L. (2019) Building sustainable tourism hierarchical framework: Coordinated triple bottom line approach in linguistic preferences, *Journal of Cleaner Production*, **229**, 157-168.

4 'A New Narrative of Place': Neolocalism and Heritage-based Tourism

Michael Fagence

Place-attachment can be strengthened through storytelling and a heightened consciousness of local history (Schnell and Reece, 2003: 57)

Abstract

It is the purpose of this chapter to experiment with 'a new narrative of place' by exposing 'nodes of geographical meaning' and to explore the action of telling a story through the medium of heritage-based tourism. Using particular episodes from the history of Australia as a case study, the intention has been to unlock some of the mysteries and potentials of places and 'things' and their contributions to a distinctive neo-local identity and purpose. To do this, this study has drawn into the discovery process a number of conceptualizations and methods from the disciplinary toolkits of geography and semiotics. This approach, whilst focusing attention on a particular form of commercial neolocalism with its dependence on a special 'sense of place' and associations with that place, has been designed to expose the opportunities of linking neolocalism studies to the expansive domains of tourism studies, heritage studies and studies in regional science.

Introduction

Although commentaries about the phenomenon of neolocalism have been active for several decades, and the portfolio of case studies continues to grow, one of the exciting aspects of this area of study is that it has the potential to provide a portal of access to many companion areas of study. It always, so it seems, has 'sense of place' as a core focus. Whether the scholarship is concerned with the fundamentals of neolocalism per se, or with any one of many possible case studies of fitting commercial or industrial activity, or community development, or lifestyle preferences to those fundamentals, the end-product can often challenge conceptualisations about 'sense of place' and how it can be managed so that those fundamentals are impacted advantageously. This is a study area where, among others, there are perspectives from aesthetics, culture, economics, environment, geography, politics, psychology, and sociology that can bring enlightenment to the processes of appreciating, understanding, creating, and sustaining 'sense of place'; and it is in this expansive context of opportunity that this chapter is set.

The purpose here is to consider the nature and scope of neolocalism from a perspective and position that is somewhat different from what may be described as 'mainstream', and because of this, it may be interpreted as being explorational and experimental. Three issues are used as anchor points for the exploration. One of these has a link to neolocalism through the speculative propositions from Schnell (2013) about 'a new narrative of place' and from Ryden (1993) about 'nodes of geographical meaning'. The aim of this study is to consider what those propositions offer for the study of neolocalism. A second anchor point particularises the subject matter of the case study. Telling a story from history through the medium of heritage-based tourism is the focal point; it will be an example of commercial neolocalism. The third anchor point is concerned particularly with methodological issues, and its operational focus is to engage with a form of multi-methods research in order to expose and refine the usefulness of the body of evidence that tells a story through tourism. By working through these three issues, the over-arching aspiration is to nudge the profile of neolocalism studies towards a frontier that focuses on 'commercial neolocalism' and, more ambitiously, towards the expansive domains of tourism studies, to heritage studies, and even towards studies in regional science. Although approached in this chapter as a single task, it almost amounts to three experiments in the one study.

This might sound as though the whole exercise will be esoteric, designed to satisfy a longing for an ill-defined objective far beyond the core of neolocalism and its focus on such 'mainstream' issues as quality of life, lifestyle, environment, and 'least disturbance' from outside forces. Certainly, the drive through this chapter will be to achieve some outcomes that are pitched in research areas of methodology and even in popular culture; but the core issue will be to tell a story from the history of Australia through the medium of heritage-based tourism as a case study of commercial neolocalism. There is already a recognisable volume of case studies that loosely associate tourism with neolocalism, and these are studies that have a focus on, for example, microbrewing, retailing, antiques and books, food and drink and other local festivals, and experiencing the lifestyles of particular cultural and ethnic groups (Frank, 2018; Gibson & Connell, 2011; Holtkamp et al., 2016; Michael, 2002; Trollinger, 2012). Beyond these lie the always-inspirational commentaries of Shortridge (1989; 1996; 2005).

Among the most often-cited foundational commentaries about neolocalism are those of Flack with his description of neolocalism as an attempt 'to re-assert the distinctively local' and as a response to 'a landscape in danger of losing its uniqueness' (1997: 38), and of Shortridge with his suggestion that neolocalism is 'a deliberate seeking out of regional lore and local attachment by residents' (1996: 10). Building on more than two decades of their own experience and research, and drawing from the ponderings of many commentators, Schnell and Reese have ventured an opinion that neolocalism is the 'conscious attempt of individuals and groups to establish, rebuild, and cultivate local ties, local identities' (2014: 168). Embedded in some of the commentaries are references to such matters as identities, images, landscapes, local ties, places, place-identity, place-personality, 'sense of place' and traditions, and references to such inspirational sources, as for example, Cosgrove and Daniels (1988), Relph (1976), and Tuan (1980, 1991). Recently, Schnell and Reese have made the effort to tie many of these elements and sources together into what they have described as 'this powerful concept of neolocalism' (2014: 167).

It is from this background that the two propositions of Schnell and Ryden have been plucked to drive this study. Whereas, because of the weight of his published work, it may be claimed that Schnell is a substantial commentator to both the foundational literature on neolocalism and to its growing volume of case studies, the case of Ryden is different. His contribution comes from studies about folklore and 'sense of place'. He has suggested that places are 'fusions of experience, landscape and location' (1993: 39), are 'thickly layered with significance' (p. 95) and have appeal/attraction for both 'insiders' (i.e., those that live at the places) and for 'outsiders' (i.e., those that come to visit for reasons that might include tourism). Although Ryden's commentaries do not refer specifically to neolocalism, even a rapid read of his observations will reveal that his focus is on what landscapes have to say and tell, and how local places are concentrations of material culture, of stories, and have a personality. Schnell also addresses these matters in his proposition about a 'new narrative' (see 2013: 82-84), and complements them with observations about broadening the basic elements of neolocalism to embrace new forms, to create 'special' places and to forge positive and creative strategies.

This study uses the propositions about places by Schnell and Ryden as a starting point in the push towards the lofty aim of contributing to a new frontier of studies about neolocalism, and it draws inspiration from some of the ideas about different conceptualisations of neolocalism shared by Shortridge (1989; 1996; 2005) in his studies of rural regions across the American Mid-West. Although this study is not too far removed from the previously-mentioned case studies of microbrewing, antiques-retailing, and holding festivals, in that it has a focus on activities that are locally-important, draws on local resources, and has a conspicuous identity, there are points of difference. Among these are that the particular 'commercial manifestation' considered here is held together by threads of a story from history, that a complement of locations is needed if the integrity of the story is to be sustained (i.e., if the story is to be told in its entirety, several distinct places need to be included), that the story is an important constituent of a world-wide phenomenon, and that the telling of it through the medium of tourism is a 'leap' into a recently-established niche of popular culture often referred to as 'consuming history'.

The crux of this investigation, in short, is to test the applicability of the propositions about 'a new narrative of place' (Schnell, 2013: 84) and 'nodes of geographical meaning' (Ryden, 1993: 5). In the two sections that follow this Introduction, there is a description of the research process adopted for this study and then a commentary on a number of important interpretational and disciplinary influences. That is followed by the case study of the nineteenth century Australian bushranger-cum-outlaw Ned

Kelly. A concluding section will offer judgements about what can be learned from this experiment for the cause of research into neolocalism.

The study process

In the opening paragraph of this chapter, it was made clear that the intention was to engage in an experiment to move the study from the mainstream track of technical discussions about neolocalism and the orbit of the most frequently-used case study subjects, and, by doing that, to make progress towards a special form of 'commercial neolocalism'. A companion intention has been to conduct this study in a way that would make a contribution to the task of positioning studies about neolocalism into the established domains of tourism studies, heritage studies, and studies in regional science.

Approach to the study

Past experience of handling the body of evidence at the core of this study has proven that successful outcomes flow from adopting a model of heuristic research and drawing on perspectives and investigative and interpretive tools from the domain of the qualitative sciences. In crafting the approach to this study, a number of advocacies and challenges were especially influential. By way of examples, Moustakas' (1990) advocacy for a process that could lead to the explication of meanings and patterns of evidence, Jamal and Kim's (2005) challenge to develop well-theorised and integrated frameworks for studies positioned at the confluence of heritage and tourism, and Pearce and Butler's (2010) advocacy for a process of inquiry that could bring together conceptual and methodological dimensions from diverse study areas, all proved to be especially powerful. Similarly, the cases of Ostrom (2011) for research frameworks, van der Duim (2005) for 'scientific modes of ordering', and Creswell and Clark (2017) for mixed and multi-disciplinary methods in investigations contributed significantly to shaping the form of, and process for, this study. An important determining agent in the final sections of the case study phase was Jansen-Verbeke's call to meet the challenge of seeing, reading, interpreting, and making sense of the cultural landscape: 'we need landmarks, icons and local stories to identify the places we live in and visit ... they tell the story of a place and the habitat of many previous generations and [they] construct our identities' (2016: 279).

A synoptic explanation of the three-phase and six-step approach to this study is set out here; it should be read in conjunction with Figure

4.1. Although the process is presented as having a linear form, in practice there were recapitulations throughout the exercise, and particularly within phase 2, within phase 3 and across phases 1, 2 and 3.

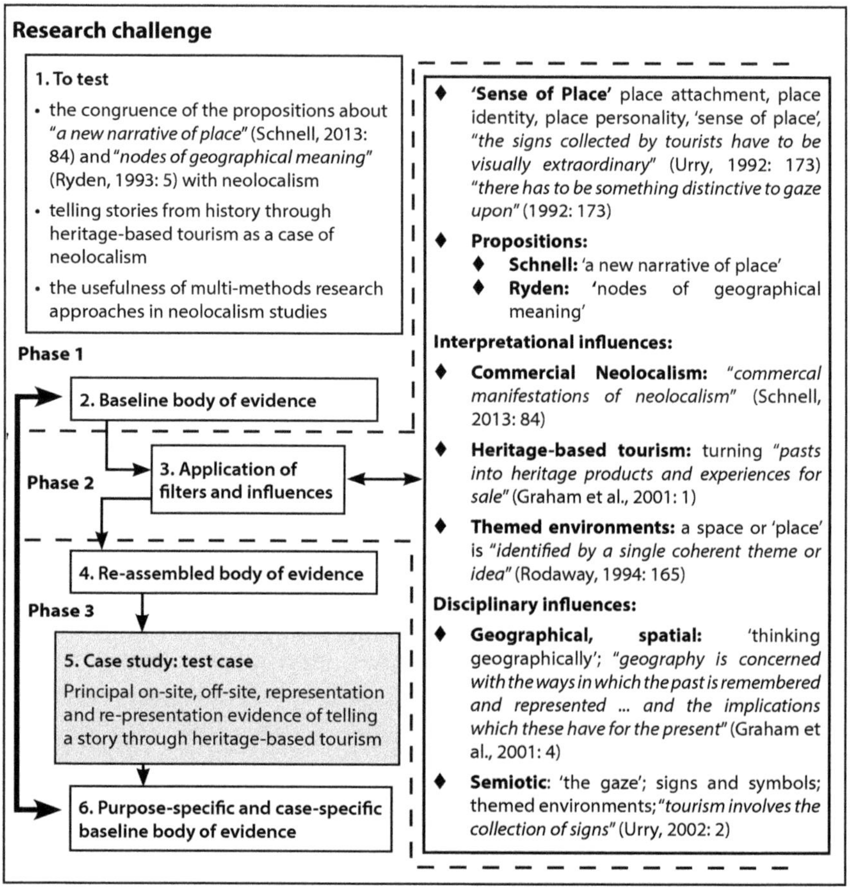

Figure 4.1: The research process.

The purpose of Phase 1 was to set out the nature and intent of the study, to nominate what was to be tested (i.e., the two propositions, the subject of the case study, and the investigative and interpretive approach and methods to be used), and to signal the establishment of a baseline body of evidence. This study has drawn on a core of evidence first assembled more than a decade ago and was given a shape that complied with van der Duim's (2005) advocacy of a framework of three components – physical and material reality, 'texts', and re-presentations. It is a core that has been maintained within that framework and has been updated through correction and augmentation from a process of ongoing desk-based and fieldwork investigations.

It is in the operation of Phase 2 that a suite of filters and influences are brought into play so as to manipulate the body of evidence into a form that is best suited for the purposes of this study. There are four sets of influences. There is the over-arching influence encapsulated in the perspectives about 'sense of place', place-attachment, place-identity, place-personality, and the importance of 'difference' and 'significance' promoted in the commentaries of Urry, in particular, and drawn from the observations of Relph and others about place and placelessness. Then, there is the important direction given to this study by the propositions of both Schnell and Ryden; together these provoke a close inspection of the advocacies of neolocalism. A third set of influences is drawn from the interpretations used in this study about the nature of 'commercial neolocalism', of heritage-based tourism and of themed environments; and the fourth set is formed from a commentary about multi-methods research in general, about epistemological pluralism in particular, and about the special contribution made in a plural form from 'thinking geographically', and from 'the gaze' element of semiotics.

In Phase 3, the purpose has been to re-assemble (or, perhaps more accurately, to rethink and reconceptualise) the body of evidence, to present the case study, and then to offer judgements about the two propositions, about the testing of them through a case study in heritage-based tourism, and about using a multi-methods investigative and interpretive approach and process.

Influences on 'sense of place'

In order to progress from what could be described as a 'raw' body of evidence about the Ned Kelly story to a state at which it would be usable for the purposes of this study, a number of filters or influences were applied to it. To summarise the filters used here, one sets the over-arching context ('sense of place'), a second defines the direction of the investigation (the two propositions), a third sets the nature of the case study (commercial neolocalism, heritage-based tourism and themed environments), and a fourth informs the methodological parameters that guide and shape the investigations and interpretations. Some elements of the fourth of these were considered in previous comments in the *Study process* section, and the implications of some aspects of the other filters will be evident through the case study section.

'Sense of place'

This is an important backgrounding influence, and it is evident throughout this study. It holds the processes of investigation and interpretation to the crucial focus on 'place' through all the considerations about neolocalism. Although there are many senses involved in creating and appreciating tourism experiences, the one that has captivated both heritage and tourism studies most is probably 'the gaze' (see later in the section on semiotics). Its importance is highlighted by the advantage it gives to tourism over other forms of presentation, re-presentation and communication, especially within the frame of popular culture. The importance 'sense of place' holds for tourism, and for any neolocal enterprise based around a tourism theme, is that the visual dimension of many circumstances signals 'difference' and 'significance', and it is these matters that will influence the degree to which resources with tourism potential may bring about a change to the status of any 'place' by raising its position from being commonplace to one that attracts serious levels of attention and attracts visits from tourists.

Propositions

The two propositions that tie this study to the circumstances of neolocalism are:

- from Schnell: a new narrative of place (2013: 83). At the point he makes this commentary about 'deliberate identities', he is making observations about 'neolocals' taking control over the forces that give shape to their communities, being creative, and resisting external abstract forces from taking control of their community's destiny. He lists as contributions to this 'new narrative' establishing a neolocal identity based on local knowledge, a local economy, local resources and connections, local narratives and stories, and by seizing local opportunities to be part of a broader economic, political, and social undertaking. Elsewhere in that commentary he makes reference to the geography of neolocalism (2013: 76-81) that provides the link with Ryden's proposition.
- from Ryden: concentrated nodes of geographical meaning (1993: 5). His entire commentary about 'mapping invisible landscapes' is focused on making 'sense of place'. For example, he comments on invisible landscapes of meanings, memories, and associations (p. 41), he refers to 'places' as fusions of experience, landscape, and

location (p. 39), and to landscapes being 'places' of objectification of the past, a catalyst for memory (p. 39), and he considers geography to have an ordering function (pp. 53-58) and being 'thickly layered with significance' (p. 95).

Linking these two propositions leads to a position in which an intention of neolocalism is for communities to seize local opportunities over which they can exert some control, and to use an ordering device to unravel, reveal, and then exploit those aspects of any history associated with those communities as a catalyst for the improvement of economic, political, and social circumstances. The case study in a later section addresses these matters.

'Commercial neolocalism'

In his commentary on manifestations of neolocalism, Schnell (2013: 56-64) refers almost exclusively to what he later describes as 'commercial manifestations of neolocalism' (p. 84). Case study evidence is dominated by a range of industries and businesses that lean heavily on local resources, production, retailing and servicing processes, and by communities that host regular festivals. Schnell's case is that communities accessing their local resources for food and for making goods is not a new phenomenon. What is new, however, is the degree to which those local communities seek to expose, exploit, and control the development of those resources and services. This is where there is a linkage with the circumstances of tourism and neolocalism – a local resource that may/may not have associations with 'a bigger picture', and that may/may not be exploited, and that may/may not have a ready-made clientele.

Bird's (2002) commentary on grappling with cultural identity and 'places', and the case studies of Shortridge on the experience of the State of Kansas, report many useful lessons. It is the nature of tourism to be exploitative, and therein lies a possibility that tourism activity could jeopardise the 'interwoven set of values and ideals' that are typically associated with neolocalism. For example, whereas the 'baseline' characteristics of neolocalism may be inward-looking, locality-supporting and locality-protecting, those of tourism are almost always outward-looking, locality-maximising and attention-seeking. Some of this potential jeopardy may be coped with by spatial solutions of layering, clustering, and networking (Jones & Alderman, 2003; Michael, 2003; Schnell, 2013).

Heritage-based tourism

The term 'heritage' has attracted a cluster of interpretations, not least because of its heterogeneity and ubiquity. By way of example, Lowenthal has claimed that 'all at once heritage is everywhere ... a prime lure of tourism. One can barely move without bumping into a heritage site' (1998: ix); while the opinion of Johnson and Thomas is that 'heritage' is virtually anything by which 'some kind of link, however tenuous or false, may be forged with the past' (1995: 170). In recent decades, in both practice and scholarship, there has been a change of focus with 'heritage' considered to be a process by which evidence from past periods has been converted into products and situations (Harvey, 2001). This approach has come to dominate thinking about the linkage of history-linked resources with tourism (de Groot, 2009; Jansen-Verbeke, 1998; Robinson & Silverman, 2015; Staiff et al., 2013). Graham et al. have taken a particular view about this with their reference to heritage-based tourism as being a process of turning 'pasts into heritage products and experiences for sale' (2000: 1). Whilst the same repertoire of history-linked resources may attract interest from such popular mediums as dance, drama, literature (novels, poems), music, painting, and sculpture, the presentation through tourism takes a distinctive form with, for example:

- Direct engagement with the physical, tangible and empirical features of history (buildings, diaries, documents, artefacts, and tools);
- Visits to witness and experience sites, buildings, and constructions;
- Visits to view and inspect items in galleries, libraries, and museums;
- Encounters with the transcendental and less tangible aspects of history communicated through folktales, folklore, legends, and myths; and
- Engagements with commentaries, interpretations, and opinions about historic events and the actions of people in past periods.

Themed environments

Heritage-based tourism can be manifested at single places, or in groups of places, or it can be dispersed across small regions. In each of these cases, it will often be a theme related to a particular happening or event or person that causes it to be 'different' and 'significant'. Rodaway has captured these ideas of difference and significance in his conceptualisation of a themed environment or themescape that he describes as a space of 'place' identifiable by a single coherent theme or idea (1994:

165). Common examples of this phenomenon in tourism would include, for example, battlefield sites, past residences of historically-significant figures, antique-retailing districts, heritage precincts, or clusters of places that share in telling parts of a story. For each of these, the previously-mentioned backgrounding issues of 'sense of place' will be a distinctive influence. In the case of story-telling through tourism, the 'canvas' for the story may exceed a single site, and may extend to a number of (not necessarily contiguous) sites within a single township, or even across a rural area composed of many small townships. For this more spatially-extensive situation, the most likely configuration for telling the story might be networks or clusters as the various sites are drawn in to share their particular contributions. It is for this situation that Ryden's proposition about 'concentrated nodes of geographical meaning' is especially appropriate.

Methodological issues

Accumulated evidence from a selection of case studies of 'commercial neolocalism' tends to support comments made earlier in this chapter, that the dominant issues and perspectives are economic and social. Another commonality is that most studies draw on the approach, perspective, and toolkit of a single discipline in both the investigations and the interpretations. Opportunities to develop deeper and wider levels of understanding about neolocalism may be missed if this situation persists. In recent decades, there has been a rapidly-accumulating body of literature about the desirability of, and the usefulness of, combinations of study approaches that can lead to more thorough levels of understanding in research; for example, sources on multi-methods and mixed methods research, such as Creswell and Clark (2017), and on interdisciplinary research approaches by Darbellay (2016). This is one of the challenges accepted in the experimental approach taken in this chapter.

For this study, there has been engagement with a research process of dialectical pluralism (Johnson, 2017) and epistemological pluralism (Suri, 2013). The purpose of this has been to engage with an investigative and interpretive approach that would cope well with linking together what has happened (history) with where it happened (geography) and the signs and symbols of the what and the where (semiotics). The approach taken here recognises and accommodates that, and it achieves it by engaging with the multiple and overlapping descriptions from the previous singular domains and foci (i.e., the what, the where, and the symbolism of both), and reformulating them into new descriptions that are particularly appropriate for coping with the matters of distinctiveness, visibility, association, and relevance that are generally considered to be so important for creating and sustaining interest in tourism (Urry, 2002). These were among the principal considerations that drew this study into a pluralism-configured research process that draws on compatible elements from the disciplines of geography and semiotics.

Shurmer-Smith's explanation of geography is that it is 'the field of study which concentrates upon ways in which space, place, and the environment participate in an unfolding dialogue of meaning ... [including] ways in which place, space and environment are perceived and represented, and how they are depicted' (2002: 3). The remit of geography has come to focus on the distinctiveness of places, cultures, economies, histories, landscapes, and organisational systems, in both in situ situations and in various inter-relationships. There is a determinable way of 'thinking geographically' (Hubbard et al., 2002) that bridges the observable world with the abstract world of theories, ideas, and propositions. At its core, both empirically and methodologically, the principal issues dealt with are concerned with functionality, location, 'sense of place', significance of distance, space and place perceptions, and spatial patterns. The companion disciplinary area in this study has been semiotics, and specifically 'the gaze' component. Commentators have lauded semiotics as a research methodology because of its versatility and durability, and because of its never-ending search for meaning. Sless (1986), for example, has referred to it as an exercise of 'intellectual curiosity', and opportunities of using semiotic perspectives in studies in tourism have been explored by Metro-Roland (2009), and in heritage studies by Waterton and Watson (2014). For Urry (1990, 1992, 2002), semiotics has been championed as a process for identifying distinctiveness. For this particular study, being able to identify the distinctiveness of places, buildings, structures, artefacts, documents, artworks, and so on with their linkages to the story being told about Ned Kelly has proved to be crucial to interpreting the contribution of all of those elements to that story through layers of meaning and through various spatial structures and linkages. The impact of these various influences is considered through the case study.

Case study

The subject matter of the case selected for this study is unlike many of those that have been referred to in previous paragraphs. It takes issues of neolocalism further into the realm of tourism than other studies, and it does this by concentrating the focus on the special domain of telling stories from history through tourism. This sets the case study into the scope of history-linked and heritage-based tourism, and it exposes a number of circumstances that have the potential to change a small township or 'place' from one form, function and identity to another, and, with this impact, change the expression of neolocalism. This case conforms with what Jansen-Verbeke (1998) has referred to as an expression of 'tourismification' in which a township or 'place' draws on localised resources that support the telling of a story from history through the medium of tourism. In this case study, the story is about the nineteenth century Australian bushranger-cum-outlaw Ned Kelly. In the telling of it, particular attention is drawn to those tangible and intangible elements and resources that have impacted on how the townships and 'places' in a rural area in the State of Victoria in Australia have been drawn into 'product and experiences for sale' (Graham et al., 2000).

The story of Ned Kelly

The first task is to tell the story of Ned Kelly briefly. As a backgrounding comment, it is worth noting that the story of Ned Kelly is a good fit with narratives about the lifestyles and lawless activities of a group commonly referred to as bandits, highwaymen, outlaws, and others. The actions of this group were important contributions to the evolving cultural and political history of parts of Europe, North America, and Australia in the eighteenth and nineteenth centuries (Hobsbawm, 1969). For the history of Australia, the case of Ned Kelly is of such prominence that Seal has referred to it as 'a compelling example of the potency of the outlaw hero tradition ... [with Kelly] the closest thing Australia has to a national hero' (1996: 145). For the purposes of this study, there are four important elements to the story: when it took place, where it took place, a sequence of dramatic events, and the legacy that remains. In summary, the story of Ned Kelly's life spanned the period 1855 to 1880, but it was the dramatic events of the period 1878-1880 (see later) that have hallmarked its special contribution to the history of Australia.

Members of the Kelly family were caught up in a momentum of lawlessness as many of the rural poor responded, sometimes violently, to a range of issues that they considered discriminated against them. The targets of the rural outrage were the banks, some owners of large estates, the police, and the judiciary. Kelly's responses invested him with a special profile and notoriety, and late in the period of the dramatic events, he prepared a political manifesto that aired the grievances of the rural poor and sought the creation of a separate State. The dramatic events, often referred to as the 'Kelly Outbreak' (McQuilton, 1979), were escalations of the general lawlessness, and included a siege, the murder of three policemen, two bank robberies, the murder of a police informant, and a conclusive siege at which three members of the Kelly gang were killed and Ned Kelly captured. The final step in the drama was Kelly's conviction and execution in Melbourne Gaol in November 1880. The geographical 'footprint' of the action in the Kelly Story took place within a loosely-defined 'corridor' set in the north-east region of the State of Victoria. It straddled the principal Sydney-Melbourne route, and it boasted a pattern of many small townships and isolated farmsteads. There is energy from two opposing strands of thought about the happenings in the Kelly story that contribute to the persistent telling of it, not only through scholarship, but also through various forms of popular culture, and especially through heritage-based tourism.

Importantly, for the purposes of tourism, the legacy of the Kelly story is 'captured' in such evidence as, for example (see also Tables 4.1 and 4.2):

- Buildings, sites of buildings and events, structures, camp sites and sites of both Kelly Gang and police look-outs;
- 'Text' forms, including gravesites, memorials, and monuments;
- Kelly-related personal items including his guns and rifles, the suit of armour he designed for the final siege, his personal political manifestos (the Cameron and Jerilderie Letters);
- 'Official' documentation, including parliamentary reports, Commission of Inquiry reports, court proceedings, gaol records; and
- Photographs, newspaper, and magazine reports.

Much of this evidence is accessible and can be 'experienced' on-site or near to the location of the story's events, and some of that which is off-site is located in special collections in art galleries, libraries, museums, and public record offices, and still more is accessible by means of re-presentations of the story in film, drama, artworks of various forms, music, and many forms of literature.

Table 4.1: Primary categories of evidence

Physical and material reality: buildings, places, sites, settings

- Domestic buildings (homesteads)
- Public buildings (courthouses, gaols, holding cells, telegraph offices)
- Structures (bridges)
- Tracks, trails
- Look-outs, camp sites, caves
- Sites of bank robberies
- Sites of sieges
- Townships

Texts

- Artefacts, tools, appliances
- Armour, weapons,
- Gravesites, memorials, monuments
- Bravery award
- Death mask
- Diaries, personal documents and letters
- Records of Commission of Inquiry, parliamentary records, court transcripts, gaol records
- Photographs, newspaper, and magazine reports

Presentation, re-presentation

- Items from **texts** held in museums, public records office, libraries
- Items in art galleries
- Commemorative re-enactments
- Animatronic displays
- Reconstructions
- Collections, memorabilia
- Use of story-related symbols in commercial products, road signs, advertisements

Narratives of place; nodes of geographical meaning

Using the facts and the circumstances of the Ned Kelly story as the focal point and being able to access many of the resources at, or in close proximity to, the 'places' where the events occurred gives to the telling of the story both geographical and semiotics-based realism. Much of the evidence of the legacy of the story is still accessible, and in many cases can be seen and touched. Many of the significant items, together with their locations, are listed in Table 4.2 (see also, Figure 4.2).

Table 4.2: Primary 'symbols' of the Kelly story

Physical and material reality: buildings, places, sites, settings

- Remnants, homesteads at Avenal, Beveridge and Greta
- Bushranger look-outs, camp sites, caves especially in Woodshed Valley
- Sites of sieges at Glenrowan, Stringybark Creek
- Prison holding cells at Beechworth, Benalla, Melbourne
- Courthouses at Beechworth, Benalla, Mansfield, Melbourne
- Hughes Creek, Avenal (rescue of boy by Kelly)
- Gaols at Beechworth, Melbourne
- Sites of bank robberies at Euroa, Jerilderie

Texts

- Gravesites at Avenal, Beechworth, Benalla, Greta, Mansfield, Melbourne
- Memorials, monuments at Glenrowan, Mansfield, Stringybark Creek
- Suits of armour worn by Kelly gang
- Cameron, Jerilderie Letters
- Kelly sash (for Avenal rescue)
- Death mask
- Diaries
- Records of Commission of Inquiry, parliamentary records, court transcripts, gaol records
- Photographs, newspaper, and magazine reports

Presentation, Re-location, Re-presentation

- Items from texts held in museums, public records office, libraries in Beechworth, Benalla, Melbourne
- Items in art galleries – Nolan's Kelly series in Canberra, Nolan's tapestry and Tucker's portrait in Benalla
- Re-enactments
- Animatronic display (Glenrowan)
- Reconstructions (Beechworth, Benalla, Glenrowan, Melbourne)
- Collections of 'Kellyana'
- Uses of Kelly armour symbol in commercial products, public notices
- Re-presentations of the story in drama, film, literature, music

That evidence can be fitted to four categories as follows:

- 'On-site' evidence:
  - ◊ Sites such as bushranger/outlaw lookouts, camp sites, cave sites, siege sites, bank robbery sites;

  - ◊ Townships, remnant townships;
  - ◊ Physical and material objects, including courthouses, gaols/prisons, banks, residences, inns, telegraph offices, prison lockups, railway stations, and constructions such as bridges, railway tracks, and stations;
  - ◊ Monuments, memorials.
- 'On-site' remnant or indicative evidence:
  - ◊ Remnant townships (especially abandoned gold mining camp sites);
  - ◊ Remnant homesteads (of the Kelly family, and of the pre-bank robbery 'bail-up' homesteads);
  - ◊ Unmarked gravesites;
  - ◊ Plaques commemorating actions (such as the bank robberies).
- 'Off-site' evidence in collections in art galleries, libraries, museums:
  - ◊ The Kelly Gang armour and armaments;
  - ◊ Kelly bravery sash;
  - ◊ Kelly death mask;
  - ◊ Documentation, including Commission of Inquiry reports, parliamentary reports, court proceedings, newspaper and magazine items, personal letters, photographs, personal political manifestos and letters.
- Evidence from forms of representing or re-presenting aspects of or items from the story:
  - ◊ In film (eleven), in theatre and drama (seven), in television miniseries (one), in musical theatre (one);
  - ◊ Re-enactments in festivals, commemorative events.

While this is only an indicative list of evidence, it is clear that the range of evidence for telling stories through the medium of tourism is most likely more extensive than is possible for the more-narrowly focused accounts of neolocal industrial, commercial, and festival processes. However, with this weight of evidence, the challenge of 'display' for the purpose of tourism encounters two practical difficulties:

- A geographical challenge; and
- A semiotic challenge.

Although telling a story may be locationally-fixed to one 'place', it is in the nature of stories to 'roam' across a number of 'places'. As the 'sense of place', place-attachment, place-identity, and place-personality may migrate as the telling of the story unfolds, tourism will adjust to that by being conducted through the medium of a geographical structure that is 'layered' to recognise the relative importance of different spaces located at the same 'place', or is multi-nodal as a network of linked 'places' of similar importance. In both cases, the tactic might well be to recognise differentiated levels and styles of themed environments and for telling the story through a combination of spot locations, street-based ribbons, precincts, clusters, and districts. In the case of the Ned Kelly story, there are examples of each of these (see Figure 4.2 and Table 4.2):

- 'Spot' locations: a remnant Kelly homestead (Beveridge), the residence of a principal contributor to Kelly court cases (Beechworth), the site of a siege (Stringybark Creek), a cemetery (Greta), a look-out, (Harry Power's Look-out);
- A dispersed pattern of isolated 'spot' locations within a township: Courthouse, gravesites, and a memorial (Mansfield); a museum and gallery of artifacts and art works (Benalla); a siege site, a re-constructed railway station, and a number of memorabilia retail outlets (Glenrowan);
- A 'ribbon' configuration within a township: Telegraph office, blacksmith's shop, and others that share aspects of the story being told (Jerilderie); and
- A defined heritage precinct within a township with several contiguous buildings and sites that share the telling of a particular episode of the story (Beechworth).

The locations of the body of evidence for this story have some consistency with Ryden's proposition about 'concentrated nodes of geographical meaning', Schnell's (2013) ideas about networks, and Michael's (2003) articulation of a form consisting of clusters and nodes in a network. Figure 2 provides evidence of the geographical patterning that underpins the telling of the Ned Kelly story. In using the Ned Kelly story as a case study, it has been possible to expose how the inter-meshing of a number of strands of thought can contribute to telling a story from history, and how that story can respond to those influences and be expressed as a form of neolocalism.

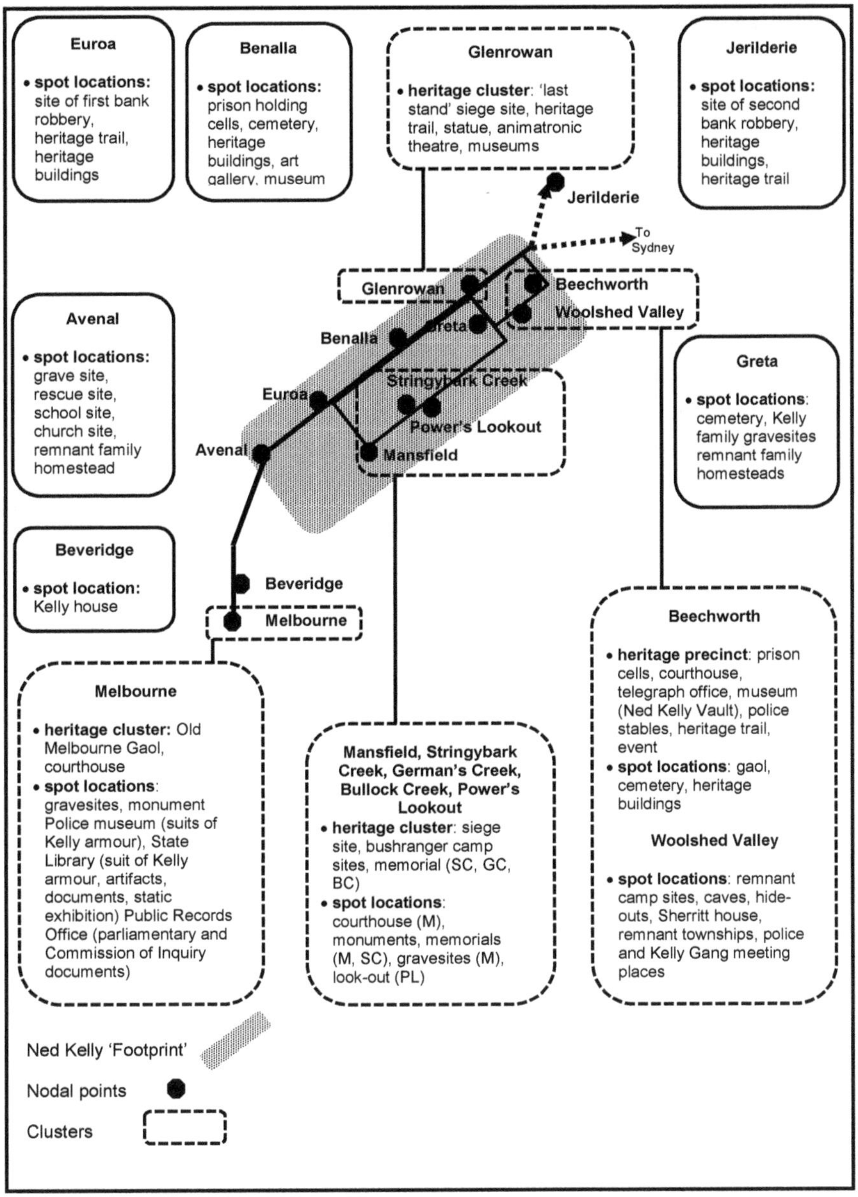

Figure 4.2: Nodes of the Ned Kelly story.

Conclusion

In the Introduction, reference was made to the opportunities for using studies of neolocalism as a portal to provide access to companion areas of study. This was not intended to suggest that studies having neolocalism as their core will necessarily open up access to other study areas, but that, as a portal does, it would provide a pathway for connections to be made between study areas and to the disciplines that drive them, and especially to provide opportunities to take advantage of theories and practices from those other areas of study to improve the situation and capacity of studies of neolocalism. Because focusing on heritage-based tourism (and even a niche component of that) was not intended to be a straightforward substitution for previous case studies of neolocalism, this study set out to engage with a process of explication by focusing attention on two propositions (from Schnell and Ryden) that together offered an opportunity to make an assessment of, not only, the basic case of neolocalism as 'a narrative of place' but also the geographical expression of it as 'a new narrative'. Just as the subject of the case study was different from 'mainstream', this investigation sought to draw help through the portal of opportunities from a configuration of disciplines that was also different from 'mainstream'. Whether or not this tactic of making even a small change to the subject matter, and the investigative and interpretive processes commonly used in studies of neolocalism, will achieve for those studies access into the domains of tourism studies, heritage studies, and to studies in regional science can only be determined over time. The work of this case study is a step in that direction.

What has been achieved? The brave statements of the opening paragraphs have been addressed. The selected case study, even though outside of the loose parameters of 'mainstream', has demonstrated that the concerns of neolocalism are significant in situations where the local resources have potential for tourism development. It has also revealed that, where telling a story from history is the focus of that development, restricting the impact of that to a single 'place' may be unlikely, and in that situation, issues of 'sense of place' will be confronted by challenges that may need to draw on disciplinary kit-bags that can unlock some of the mysteries and potentials of places and things with previously-overlooked tools of investigation and interpretation. In this study, that engagement was with 'thinking geographically' and 'the gaze' components of semiotics. One of the reasons for giving attention to the various influences and filters in Phase 2 of the study process was to provide some insight into how other non-'mainstream' subjects might be managed if there is a need to work outside the more conventional scopes of disciplines, such as economics.

This study was experimental. Because of the special nature of the selected case, it is unlikely that the outcomes will have a general applicability across all cases of neolocalism. However, the approach taken and the methods used in this study offer insights into opportunities to complement the common repertoire used in studies of commercial neolocalism, and in so doing, to increase the levels of understanding of neolocalism as a global movement and to increase the capacity of those studies to influence the planning and development of neolocal situations.

References

Bird, E. (2002) It makes sense to us: cultural identity in local legends of place, *Journal of Contemporary Ethnography*, **31**(5), 519-547.

Cosgrove, D. & Daniels, S. (eds.) (1988) *The Iconography of Landscape*, Cambridge: Cambridge University Press.

Creswell, J.W. & Clark, V.L.P. (eds.) (2017) *Designing and Conducting Mixed Methods Research*, 3rd edn, Thousand Oaks, CA: SAGE.

Darbellay, F. (2016) From disciplinarity to postdisciplinarity: tourism studies de-disciplined, *Tourism Analysis*, **21**, 363-372.

de Groot, J. (2009) *Consuming History: Historians and Heritage in Contemporary Popular Culture*, London: Routledge.

Flack, W. (1997) American microbreweries and neolocalism: Ale-ing for a sense of place, *Journal of Cultural Geography*, **1**(2), 37-53.

Frank, J. (2018) *Regenerating Regional Culture: A Study of the International Book Town Movement*, Cham (Switzerland): Palgrave Macmillan.

Gibson, C. & Connell, J. (eds.) (2011) *Festival Places: Revitalizing Rural Australia*, Bristol: Channel View Publications.

Graham, B.J., Ashworth, G.J. & Tunbridge, J.E. (2000) *A Geography of Heritage*, London: Arnold.

Harvey, D. (2001) Heritage pasts and heritage presents: temporality, meaning and the scope of heritage studies, *International Journal of Heritage Studies*, **7**(4), 319-338.

Hobsbawm, E. (1969) *Bandits*, London: Wiedenfeld and Nicholson.

Holtkamp, C., Shelton, T., Daly, G., Hiner, CC. & Hagelman III, R.R. (2016) Assessing neolocalism in microbreweries, *Papers in Applied Geography*, **2**(1), 66-78.

Hubbard, P., Kitchen, R., Bartley, B. & Fuller, D. (2001) *Thinking Geographically,* London: Continuum.

Jamal, T. & Kim, H. (2005) Bridging the interdisciplinary divide: towards an integrated framework for heritage tourism research, *Tourist Studies,* **5**(1), 55-83.

Jansen-Verbeke, M. (1998) Tourismification of historical cities, *Annals of Tourism Research,* **25**(3), 739-742.

Jansen-Verbeke, M. (2016) Tourismification of cultural landscapes, Proceedings of TCL2016 Conference, 12-16 June, Budapest, INFOTA, 276-281.

Johnson, R.B. (2017) Dialectical Pluralism: a metaparadigm whose time has come, *Journal of Mixed Methods Research,* **11**(2), 156-173.

Johnson, P. & Thomas, B. (1995) Heritage as business, in D. Herbert (ed.), *Heritage, Tourism and Society,* London: Mansell, pp. 170-190.

Jones, K.L. & Alderman, D.H. (2003) Antiques tourism and the selling of heritage in eastern North Carolina, *The North Carolina Geographer,* **11**, 74-87.

Lowenthal, D. (1998) *The Heritage Crusade and the Spoils of History,* London: Cambridge University Press.

McQuilton, J. (1979) *The Kelly Outbreak: The Geographical Dimensions of Social Banditry,* Carlton (Victoria): Melbourne University Press.

Metro-Roland, M. (2009) Interpreting meaning: an application of Peircean semiotics to tourism, *Tourism Geographies,* **11**(2), 270-279.

Michael, E. (2002) Antiques and tourism in Australia, *Tourism Management,* **23**(2), 117-125.

Michael, E. (2003) Tourism micro-clusters, *Tourism Economics,* **9**(2), 133-145.

Moustakas, C. (1990) *Heuristic Research: Design, Methodology, and Applications,* Newbury Park: SAGE.

Ostrom, E. (2011) Background on the institutional analysis and development framework, *Policy Studies Journal,* **39**(1), 7-27.

Pearce, D.G. & Butler, R. (eds.) (2010) *Tourism Research: A 20-20 Vision,* Oxford: Goodfellow Publishing.

Relph, E. (1976) *Place and Placelessness,* London: Pion.

Robinson, M. & Silverman, H. (eds.) (2015) *Encounters with Popular Pasts: Cultural Heritage and Popular Culture,* London: Springer.

Rodaway, P. (1994) *Sensuous Geographies,* London: Routledge.

Ryden, K. (1993) *Mapping the Invisible Landscape: Folklore, Writing and the Sense of Place,* Iowa City: University of Iowa Press.

Schnell, S. (2013) Deliberate identities: becoming local in America in a global age, *Journal of Cultural Geography,* **30**(1), 55-89.

Schnell, S. & Reese, J. (2003) Microbreweries as tools of local identity, *Journal of Cultural Geography,* **21**(1), 45-69.

Schnell, S. & Reese, R. (2014) Microbreweries, place, and identity in the United States, in M. Patterson & N. Hoalst-Pullen (eds.), *The Geography of Beer*, Dordrecht: Springer, pp. 167-187.

Seal, G. (1996) *The Outlaw Legend*, Cambridge: Cambridge University Press.

Shortridge, J.R. (1989) *The Middle West: Its Meaning in American Culture*, Lawrence: University Press of Kansas.

Shortridge, J.R. (1996) Keeping tabs on Kansas: reflections on regionally based field study, *Journal of Cultural Geography*, **16**(1), 5-16.

Shortridge, J.R. (2005) Regional image and sense of place in Kansas, *Kansas History: A Journal of the Central Plains*, **28**(3), 202-219.

Shurmer-Smith. P. (2002) *Doing Cultural Geography*, London: SAGE.

Sless, D. (1986) *In Search of Semiotics*, Totawa, NJ: Barnes and Noble Books.

Staiff, R., Bushell, R. & Watson, S. (eds.) (2013) *Heritage and Tourism: Place, Encounter, Engagement*, Abingdon, Oxon: Routledge.

Suri, H. (2013) Epistemological pluralism in research synthesis methods, *International Journal of Qualitative Studies in Education*, **26**(7), 889-911.

Trollinger, S. (2012) *Selling the Amish: The Tourism of Nostalgia*, Baltimore, MD: Johns Hopkins University Press.

Tuan, Y-F. (1980) Rootedness versus sense of place, *Landscape*, **24**, 3-8.

Tuan, Y-F. (1991) Language and the making of place: A narrative-descriptive approach, *Annals of the Association of American Geographers*, **81**(4), 684-696.

Urry, J. (1990) *The Tourist Gaze: Leisure and Travel in Contemporary Societies*, London: SAGE.

Urry, J. (1992) The 'tourist gaze' revisited, *American Behavioural Scientist*, **36**(2), 172-186.

Urry, J. (2002) *The Tourist Gaze*, 2nd edn, London: SAGE.

van der Duim, V.R. (2005) Tourismscapes: an actor-network perspective on sustainable tourism development, PhD, Wageningen University, Wageningen (The Netherlands).

Waterton, E. & Watson, S. (2014) *The Semiotics of Heritage Tourism*, Bristol: Channel View Publications.

5 Neolocalism, Revitalisation and Rural Tourism Development

Jessica Faustini Aquino and Gudrun M. H. Kloes

Abstract

Small rural communities in Iceland have seen a fast-changing economy and have had to come up with different ways to become resilient to these changes. This chapter takes an ethnographic approach to the study of neolocalism to describe how the community of Húnaþing vestra saw the potential to nurture a sense of place using their natural and cultural assets to enhance the local economy and community livelihoods. We will describe how certain community initiative projects have played an active role in fostering the local identity and distinctiveness of a community, and its use as a strategy for crafting authentic tourism experiences. We will discuss examples spanning over 20 years (2000-2019) based on an accumulation of first-hand experiences and research by the authors who live in Húnaþing vestra. There is a need to further investigate neolocalism and craft production in rural tourism development, as it may give insights on how a community may feel empowered by telling the story of their connections to place, and how it may influence revitalization efforts and community resiliency.

Introduction

Local identity and distinctiveness are essential in creating authentic tourism experiences, while community engagement in tourism development is integral in authentically creating these experiences. Involving a variety of local stakeholders helps to not only establish what is unique about a community, it helps to empower a community to strengthen its own identity and enhances story-making. For example, the stories we tell about our community and the people who live there may have a meaningful impact on establishing a sense of place and enhancing community pride. Stories, created from cultural traditions and personal memories, provide the listener with a deeper understanding of the storyteller's life and have been shown to exert significant influence on destination image and destination choice. This understanding is an important part of the visitor´s experiences (Moscardo, 2010), as they want to know the stories of the places they visit. Not only do visitors use storytelling as a way to consolidate and describe their experiences to others (Tung & Ritchie, 2011), these stories are created by the interactions visitors have with other people they meet, i.e., local staff (Moscardo, 2015).

Adger defines social resilience as 'the ability of groups or communities to cope with external stresses and disturbances as a result of social, political, and environmental change' (2000: 347). The ability to transform or adjust to external stressors has been linked with innovation and coping strategies that have led to community resiliency (Kokorsch, 2017). The resilience and innovation of Icelandic coastal communities, like Húnaþing vestra, have created a remarkable story about their ability to cope with external stresses, such as socio-economic, environmental, and demographic changes. Coupled with the geographical distinctness of North Iceland, it is easy to see why so many people travel to experience this story for themselves. The goal of this chapter is to communicate key milestones and events and to describe community-based tourism projects that have impacted the community of Húnaþing vestra. Two case studies will be used to highlight key examples of tourism projects (both current and past), to describe the history of tourism development in Húnaþing vestra over the past 20 years, and how neolocalism, although not intentionally, was used as a strategy for place-making, cultural revitalisation, and conservation efforts of local wildlife.

Húnaþing vestra, Iceland

Húnaþing vestra covers 3007 km^2 (Ráðbarður Sf., personal communication) and contains a total population of 1,181 (Statistics Iceland, n.d.). Concerns for community quality of life continue to be a topic of discussion. There are concerns about outward migration, as young individuals and families move to larger cities for work or education; thus producing a loss of human capital which may affect other community capitals (see

Emery & Flora, 2006, for a list of community capitals). Consequently, smaller rural communities in Iceland face significant challenges in terms of their capacity to cope with change. Strengthening community development programs, with a focus on entrepreneurship and innovation, in Húnaþing vestra has influenced young individuals and families to move back to Húnaþing vestra and to reinvest in their communities.

Northwest Iceland, in general, has seen a slow population decline over the years due to changing economies and other factors, such as better transportation routes. Route 1, also known as the Ring Road because it runs a ring along the Icelandic coast, was completed in 1974 creating what Valsson et al. (2013: 28) called a 'ribbon of habitation'. Most urban areas along the Ring Road have increased in population. Although the Ring Road cuts through the municipality of Húnaþing vestra, travellers on the Ring Road must turn off it to reach the three larger villages. For example, it is ten minutes from the Ring Road to Borðeyri (population around 15), two minutes to Laugarbakki (population 56), and four minutes to Hvammstangi (population 576).

In the 19th and early 20th centuries, Iceland saw an industrial revolution. After the 1970s, both Akureyri and the Capital Region of Reykjavík grew in importance as major population centres (Valsson et al., 2013). While at the same time, advancements in fishing technologies led people to relocate from remote farms to villages and towns with important harbours, which led to overexploitation of natural resources and ultimately led to the Individual Transferrable Quota (ITQ) (Kokorsch, 2017). Developed by the central government and introduced in the early 1980s, the ITQ was an effort to develop a more sustainable fishing management program by creating a quota system of particular fish species aimed at protecting fish stocks. However, after implementation of the ITQ, it was shown that small-boat fishermen and rural coastal communities were at a disadvantage to the dominant interest groups in national politics (Chambers et al., 2017) The 1990 Fisheries Management Act allowed quotas to become freely transferable with minor restrictions (Runolfsson, 1999), which led to quota rights migrating away from small rural communities (Chambers et al., 2017). Over the past 30 years, these coastal communities have seen their fisheries-dependent economies dwindle, leading to economic stress, which resulted in the loss of several community capitals.

This changing economy threatened the economic viability of small, remote communities in Iceland, like Húnaþing vestra. In the 1980s, farmers faced a crisis in the agricultural industry when the outbreak of scrapie in parts of Iceland, including the Northwest, led to the loss of entire sheep livestock (Aquino & Burns, in press). However, Iceland was simultaneously experiencing an increase in international visitors. Icelandic fishermen and farmers were accustomed to supplementing their income by working in other seasonal industries and saw tourism as a form of maintaining their way of life and staying on the farm (Aquino & Burns, in press). Rural farming families turned to tourism, which initially led to a boom in farm tourism (Jóhannesson & Huijbens, 2010).

Between the years of 2000 and 2010, Gudrun Kloes served as Húnaþing vestra's tourism officer with the Association of Municipalities of the Northwest (SSNV) and had first-hand experiences with several community-based tourism initiatives highlighted in this chapter. SSNV is an association of three municipalities, Húnaþing vestra, Austur-Húnavantssýsla, and Skagafjörður. Between 2000-2010, there was an emphasis on tourism in the area, and there were three tourism officers—i.e., one officer within each municipality. However, after 2010, there was a shift in development strategy, and the Northwest region pushed to develop their natural resources, such as geothermal heat, water and electricity, advanced technology and information technology (IT) development, food industries using local products and expertise, and innovation and research. After this shift, the number of positions for tourism officers with SSNV consolidated into one. Tourism in Húnaþing vestra has focused on nature-based tourism, such as wildlife watching (mainly seals, whales, and birds), equine tourism, farm stays, and other individual or organised outdoor experiences, such as geothermal activities (swimming pools), hiking tours, and outdoor camping.

Tourist arrivals in Iceland have grown from 488,622 in 2010 to 2.3 million in 2018 (Icelandic Tourist Board, n.d.) (see Figure 5.1), compared with the total population of just 356,991 as of January 2019 (Statistics Iceland, n.d.). With the growth in tourism numbers in the Northwest, the number of businesses also grew adding to the number of services for the tourism industry. Tourism in Húnaþing vestra, in general, has had a positive effect on the community by helping to increase the local economy, which also leads to community revitalisation. For example, although there were a few tourism and handicraft businesses before

2000, there was a boost in tourism businesses and other related services afterwards. The Icelandic Seal Center (ISC) was established in 2005, and the first tourism boat operator, Áki Boat Tours, started offering seal and bird watching trips and sea angling trips. However, operations of this boat stopped a few years later. Soon after, another tourism boat operator, Selasigling, began in 2010 with a focus on seal and other wildlife watching.

Other businesses established along the peninsula include a coffee house, Hlaðan Kaffihús, and Geitafell Seafood Restaurant, both launched in 2010. Sjávarborg Restaurant opened on the pier in 2015, and a couple of home-based businesses have emerged, such as Ceramic Gallery Litli Ós in 2001 and Valdalækur, handmade wool products, in 2017. Since 2013, accommodations built specifically for the tourism industry have increased in Húnaþing vestra. These include the Hvammstangi Cottages (in 2013), Hvammstangi Hotel (in 2014), Hotel Laugarbakki (in 2016), and the Hvammstangi Hostel (in 2017), adding to other farm stays, hotels and Airbnbs within the municipality. The growth of many businesses has been a direct effect of the successful 'the land of seals' place branding.

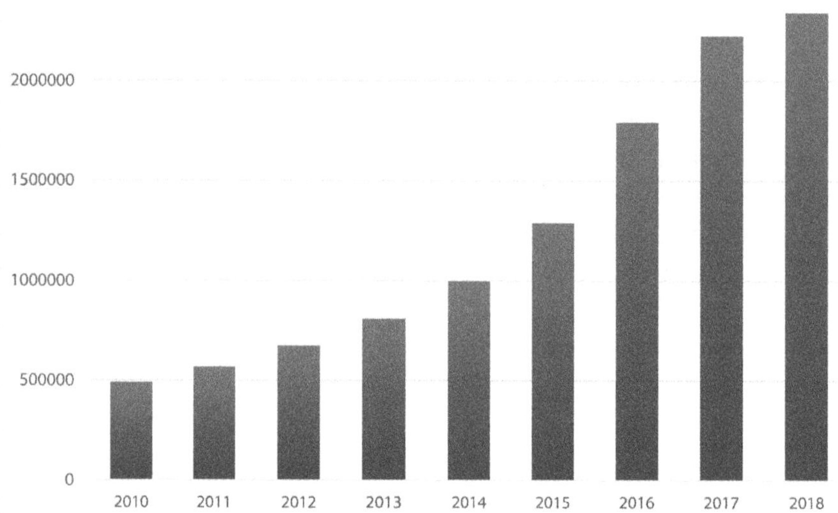

Figure 5.1: Number of foreign visitor arrivals by air and sea to Iceland from 2010-2018. *Source*: adapted by Jessica Aquino from Icelandic Tourist Board. (n.d.)

Local identity and distinctiveness

Neolocalism

Neolocalism emphasises local identity and distinctiveness. Research in the area of neolocalism has focused mainly on the areas of microbrewers and craft beer, and its use in place-making (Argent, 2018; Fletchall, 2016; Holtkamp et al., 2016). Other types of craft production may influence sense of place, such as historical and cultural stories (for instance Nordic sagas), traditional handicraft, artisan cheese, and bread. Neolocalism can be seen more broadly 'as a strong form of social embeddedness, where notions of local commodity production, place marketing, authenticity, and "wholesomeness" are folded together in support of local craft production' (Argent, 2018: 87). In this case of Húnaþing vestra, 'craft production' is the place-based tourism product. However, examining neolocalism in rural tourism development as a form of revitalisation has yet to be explored.

Eberts (2014) describes neolocalism as a marketing and branding strategy that uses geography and place to emphasise a connection to their location. This connection to a location (or place-attachment) is used in producing products for the tourism market, both locally and internationally. The pride comes from locally made products, but their consumption extends beyond the local. Tourism, more generally, has used geography and place to bridge a link to the natural and cultural assets of place. However, to use assets as a branding strategy, the brand itself must be connected somehow to a sense of place and marketed more broadly. We argue that, although the products are produced and sold locally, the marketing of these products have a greater reach in local, regional, national, and international realms within the tourism industry.

Neolocalism emphasises local identity and distinctiveness in connection with a product for consumption. For example, Flack (1997: 48) describes in his study: 'While American beer drinkers are becoming more sophisticated and passionate, their demands do not require microbreweries. There is a wide array of fine beers available from importers and larger breweries. These beers may satisfy the palate, they do not satisfy the neolocal craving.' This craving for the 'neolocal' is a way that both residents and visitors attach themselves to a place. Local communities feel a sense of pride and joy purchasing locally made products because they are familiar with the location and the person producing the item of consumption. However, the visitor travelling to a place consumes locally made products to get a sense of what it must be like to live at that place and gain a better understanding of the local culture.

There is growing literature on local beer festivals and neolocalism. A notable example include Schnell (2010), where he described the processes of neolocalism as 'a desire for, and commitment to, increased connections with place' (p. 557). Sense of place is the meaning and value people assign to a geographic location (Stedman, 2003); and 'localness', as described by Schnell (2013), is the connections between the individual, the local producer, the environment, and sense of place. As Schnell (2013) further describes, these 'local connections are seen by many as more authentic and are an increasingly important part of place-creating narratives' (Schnell, 2013: 625). In the case of Húnaþing vestra, the place-creating narratives are the cultural and natural assets of the geographic location told through crafting place-based tourism products. There is a need to further investigate neolocalism and craft production in rural tourism development, as it may give insights on how a community may feel empowered by telling the story of their connections to the place, and how it may influence revitalisation efforts and community resiliency.

Although much of the research in the area of neolocalism has focused mainly on the areas of microbrewers and craft beer (Argent, 2018; Fletchall, 2016; Holtkamp et al., 2016), less research has focused on farmers' markets and community festivals – i.e. not associated with beers – and other craft production. New areas of exploration should include the areas of handicraft production, interpretation of Nordic sagas, and other historical and cultural storytelling as a form of revitalisation.

Sense of community, community empowerment and sustainability

Sense of community is defined by Ahmad and Abu Talib (2016) as meaningfully belonging to a part of a more extensive network of relationships. The level of strength of this sense of community can vary depending upon the strength of the connection individuals feel to their community. When there are more meaningful opportunities for connecting with others through work, volunteering, or through creativity and development, it may lead to stronger connections to sense of place. Often participation is seen as a factor in strengthening the feeling of a sense of community, while local governments, policymakers and other community leaders can encourage an environment of community empowerment through actively engaging people and providing meaningful opportunities in shaping their communities.

Community empowerment is defined by Wallerstein and Berstein (1994: 142) 'as a social-action process in which individuals and groups act to gain mastery over their lives in the context of changing their social and political environment'. The concept of community empowerment has its foundations with the Brazilian educator Paulo Freire (1970) who understood that people are not just objects of action, but are actors in action, able to name their problems and their solutions for social transformation, which is a process of both individual and community empowerment. 'There is no true word that is not at the same time a praxis. Thus, to speak a true word is to transform the world, ...there is not transformation without action' (Freire, 1970: 87). Freire promoted open dialog to create a space where action could combine with reflection, which would lead to praxis and social transformation.

Community empowerment is a social process (Conger & Kanungo, 2011) and product. The process of community empowerment is through participation, seen as civil action, and the procedure is the level of support that government bodies, organisations and social groups give towards equalling out power structures. The product of community empowerment is the level of control individuals have over their own lives and their community. Therefore, community participation is critical towards both creating a sense of community, empowerment, and sustainability. As individuals of a community become more actively engaged through social interaction, it builds community capacity, which then leads to sustainable development processes at the local level.

As a community, the people of Húnaþing vestra have worked at maintaining cultural traditions and using locally grown and handmade products and projects through particular community initiatives at both grassroots and local governmental levels. A few examples of this are the Folk Museum, *Byggðasafn Húnvetninga og Strandamanna* (see www.reykjasafn.is), a museum which displays the history of the people of Húnaþing vestra, and the annual community culture and arts festival, *Eldur í Húnaþing*, where the workforce for the production and development of this festival is nearly all volunteer. The Gallerí Bardúsa is a cooperative handicraft market and merchant museum, which started in the 1990s. All of the products in this market are made locally, and women make the majority of these products. Other examples include the Grettir the Strong Project and the Icelandic Seal Center. There are also examples of local entrepreneurs that focus on highlighting cultural traditions and locally grown/made products such as, Kidka Wool Factory Shop (see www.kidka.com), Dæli Holiday Farm (see www.daeli.is), and Stóra-Ásgeirsá Farm Zoo and Horse Rental, among a few others. Because of the need for brevity in book chapters, we will focus on only two cases, the Grettir the Strong Project (GSP) and the Icelandic Seal Center (ISC).

Grettir the Strong and the Grettir Festival, Bjarg and Laugarbakki

> "He has been called the Last Viking, born on the cusp of Paganism and Christianity in the northern part of Iceland. An outcast, a fighter—a loner who was stronger than anyone around and had nothing in common with his contemporaries."
>
> The Saga of Grettir (Destination Viking Sagalands, 2005: 73)

Image 5.1: Memorial at Bjarg in Miðfjörður commemorating Grettir's mother, Ásdís. Photo by Jessica Aquino, 2019.

Gréttir Ásmundarson, also known as Grettir the Strong, was a man born in an area now known as Húnaþing vestra in 996 AD. The Saga of Grettir, written in the early 14th century (1320-1350), takes place mostly in North-west Iceland and is one of the best known Icelandic sagas retold through storytellers, plays, and poems in Iceland, Scotland, Faroe Islands, and Germany (Destination Viking Sagalands, 2005). The farm Bjarg, located in Miðfjörður, is an essential part in the Saga of Grettir. It is where Grettir was born and raised, and after his execution, it is the location where his head is buried. Grettir spent 19 years in exile, giving him the title of the longest anyone has survived as an outlaw in Iceland. Much of his story retold in the Sagas takes place at Bjarg (Image 5.1).

The Grettir the Strong Project (GSP) began in 1999 as a community initiative in conjunction with Húnaþing vestra and private stakeholders, which formed a non-profit cultural association called Grettistak in 2002. The aim of Grettistak and the GSP was preserving the culture and history of Húnaþing vestra through the Saga of Grettir and other local stories. From 1996-2010, an annual Grettir Festival in Bjarg was popular with both residents and tourists. The first festival started in 1996 during the 1000th anniversary of Grettir the Strong's birth. In the first few years, the festival was a small-scale event consisting of poetry recitals, choir singing, and people dressed in national costumes and traditional clothing. After the GSP initiative, the festival grew in scale with the addition of other activities. Festival attendees were offered guided tours of Grettir's birthplace and childhood home, with other events that included learning about the significance of historical place names and other local sites central to the Saga of Grettir (Destination Viking Sagalands, 2005). Also, the festival featured a strongman competition where both men and women competed for the Grettir Cup, games for children, and a dance which marked the end of the festival.

The GSP's first steps as a project were an inventory and report conducted by a local anthropologist, Jón Jónsson. The final report laid out the essential work and vision of the overall project. At the same time, similar saga projects sprang up at various points in Iceland. For example, the Saga Center, located in Hvolsvöllur (a small town in the South of Iceland) has an exhibition displaying the Saga of Njál (see http://www.sagatrail.is/en/museums/saga-centre-in-hvolsvollur/); as well as the Settlement Center in Borgarnes (see http://www.landnam.is/eng/). Cooperation of several Icelandic saga projects led to the foundation of a multinational cooperation under the title Destination Viking – Sagalands, between Iceland, Sweden, Greenland, Norway, Faroe Islands, and the United Kingdom (see https://www.keep.eu/project/1160/destination-viking-saga-lands for a listing of all project partners). One outcome of Destination Viking—Sagalands, which partially funded the GSP initiative, was the foundation of the Destination Viking Association (see https://www.followthevikings.com/about/destination-viking-association). Nevertheless, the most important patrons of the GSP were residents of Húnaþing vestra, the non-profit organization Grettistak, and various local and national funding sources.

Overall, the GSP was a multi-level initiative split into two phases. Phase one consisted of the development of a park in the village of Laugarbakki created for outdoor activities, such as tests of physical strength (i.e., Grettir's Cup during the Grettir Festival), archery, storytelling in the stone circle, and interpretation of the sagas and local culture. These activities were consistent with the Saga of Grettir and were in line with how Grettir battled with both natural and supernatural elements. In 2001, the annual Grettir Festival moved from Bjarg to the new park in Laugarbakki, to which Viking interpreter groups from other parts of Iceland travelled and re-enacted Viking battels or presented crafting techniques, such as metal forging, weaving, dyeing, leatherwork, and other traditions.

Phase two was the acquisition and expansion of a former storage garage next to Grettir's Park. This building served as the location for GSP's main office, cultural centre, and exhibition space for the Saga of Grettir in the village of Laugarbakki. However, this step was not completed. After the financial crisis of 2008, and due to lack of funding, the project went through a period of re-evaluation and reform. Both the village of Laugarbakki and the GSP began to search for a new 'identity'. Ideas began to emerge from the residents to plan a new hotel (Hotel Laugarbakki which opened in 2016), and a local farmer's market, *Spes Sveitamarkaður* (Spes Country Market), focusing on local food and handicrafts made using traditional techniques. Both ideas were detached from the original GSP initiative. The market opened in the summer of 2009 at the Grettir the Strong Cultural Center, with up to 50 local micro-manufacturers, and offered courses in craft techniques and food processing. The Spes Country Market workforce included volunteers who were interested in the project's success, one seasonal part-time summer staff, and one year-round part-time staff. In 2010, the bank repossessed the building because of high debt and virtually no revenue from either the Spes Country Market or the Grettir the Strong Exhibit and Cultural Center, which ended the community initiatives (including the Grettir Festival).

What is left?

Property in the park still has the stone circle created for storytelling and the 5-meter-high Viking sword, which was the landmark of the project. The municipality of Húnaþing vestra owns the grounds of the park, and the building which housed the former Grettir the Strong Exhibit and Cultural Center is now privately owned. As an integration project for cultural traditions, local stories, and the preservation of Icelandic sagas, the original initiative has, unfortunately, failed based on external influences, such as Iceland's financial collapse, as well as internal factors, such as a lack of resiliency (or sustaining the 'long breath') that is indispensable for such projects. Many people who were passionate about the GSP were left mournful 10 years after the failure of the project and are reluctant to participate in discussions about starting a similar project in the future.

The GSP had several elements of neolocalism (Table 5.1). First developed as a strategy for economic development by attracting a higher number of tourists to Húnaþing vestra, the long-term goal was to contribute to the rural development of the region. The main aims of the GSP were the preservation of local history and cultural revitalisation through the Saga of Grettir and, in doing so, the GSP was successful in developing a sense of place and local identity still felt today. For example, nine years after the GSP stopped functioning, people in the community of Húnaþing vestra still use parts of the Saga of Grettir to describe their geographical location, and his stories are still used for tourism development. The GSP was a community-based initiative that involved the municipality of Húnaþing vestra and the local non-profit organisation Grettistak. Although the GSP initially received some funding from Destination Viking-Sagalands' grant, the biggest supporters were from the local community and Grettistak, which underlines how community engagement can lead to empowerment which leads to greater local support.

Table 5.1: Neolocal elements of the Grettir the Strong Project

- ☐ Rural development through tourism
- ☐ Preservation of local history
- ☐ Culture revitalisation through the Saga of Grettir
- ☐ Tourism as economic revitalisation
- ☐ Community engagement and empowerment
- ☐ Sense of place and local identity
- ☐ Use and preservation of local knowledge

In the next section, we explore how The Icelandic Seal Center (ICS) has also played an integral role in nurturing the local identity and distinctiveness of a community, strengthening sustainable rural tourism development and empowering people at the local level to develop policies for the protection of natural and cultural resources. Seals and seal-watching tourism were seen as an asset and as a creative way to help build the tourism industry in Húnaþing vestra. Seals were used as a way of branding and as a marketing strategy for the geographic area, particularly in the Vatnsnes peninsula.

The Icelandic Seal Center, Hvammstangi

'Mér er um og ó,	'I am not at ease,
sjö börn á landi	seven children on land,
og sjö börn í sjó.	and seven children in the sea.
Ég þrái að synda frjáls.	I long to swim free.
Ég er föst á landi,	I am trapped on land,
um og ó,	I am not at ease,
sjö í sjó.'	seven in the sea.'

From: ***Seal Woman***, a traditional Norse folktale (Björnsdóttir, Sigurbjörnsson, & Kristjánsdóttir 2018: 35, English translation by authors).

Similarly to the GSP, the ISC started as a community initiative aimed at developing sustainable tourism in the Northwest region but with a focus on seals and nature-based tourism. Throughout Icelandic history, seal-hunting was a significant part of its culture. Seals were valuable for their meat, blubber, and fur, and women were considered exceptionally skilful at hunting them (Willson, 2016). Seal oil, along with shark oil, was thought to be the best for indoor oil lamps. Seal fat was used to grease leather (making it water repellent), for frying foods, and as a replacement for butter (Byock, 2001). Seals were also used in boat making, as the blubber was used to caulk ships—this 'seal tar' made the boats both waterproof and resistant to wood-boring sea worms (Byock, 2001).

Seal-hunting was important during periods of famine. For example, during the 'little ice age', as livestock farming failed, Icelanders turned more to fishing and seal-hunting. Seals are also part of traditional stories and poems. Ólafsson and Pálsson (1805) wrote about how, in the early 1800s, Icelanders both venerated seals and avoided them because of their resemblance to humans. Some of the best-known folktales in the Faroe Islands, Scotland, and Iceland are about Norse mermaids, also known as silkies or seal woman (*selakona*). Seal women, in the Icelandic folktale, are seals who, during a particular time, take off their skin to reveal a human form underneath, and sing and dance until dawn when they return to their coats and to the sea.

Since the 1960s and 1970s, visitors have been coming to the Vatnsnes peninsula for seal watching. There are several locations along the peninsula that allow easy access to harbour seal colonies. Gudrun Kloes, who worked for SSNV as the tourism officer for Húnaþing vestra, initiated meetings between 2003-2004 with the locals in Húnaþing vestra. Some of the initial meetings were to discuss local assets and developing Húnaþing vestra as a tourism destination. At first, the community did not recognise their assets. However, they did see the potential for tourism in the area and how it could lead to regional development (Aquino & Burns, in press). From these meetings, an idea was created about developing seal watching and nature-based tourism. Then, the concept of starting a non-profit organisation dedicated to developing wildlife-watching tourism with a focus on seals, called the Icelandic Seal Center, was created. Seal watching had already become popular on the Vatnsnes peninsula; however, the aim was to develop tourism in the region. In the beginning stages of the ISC, some funding was allocated to build walking trails to seal watching sites, such as Svalbarð, Illugastaðir, and Hvítserkur. In addition to developing tourism, another main aim of the ISC was empowering the community, contributing positively to economic development, and seal research.

The ISC was formally established in 2005 as a community-owned, non-profit organisation consisting of 89 individuals, companies, and organisations, which was about 8% of the population of the municipality at the time. It began operations as a small seal museum with just one part-time position. The first site of the ICS included a museum and offices and was located in the historical building Sigurður Pálmason in Hvammstangi. Renovation goals of the building were to bring out the historical character, remembered by most of the inhabitants of the elder generation, which allowed them to connect easily to the new concept. The museum had exhibition space with some outdoor interpretative sculptures about seals and other marine wildlife, in addition to an outdoor display of the traditional fish drying method called *trönur*. The first floor of the building was used as offices and research. Once at the ISC, visitors could either visit the Seal Museum and learn how Icelanders have interacted with seals over time, or they could ask for directions to designated seal-watching spots or other areas of visitor interest in the region (Figure 5.2).

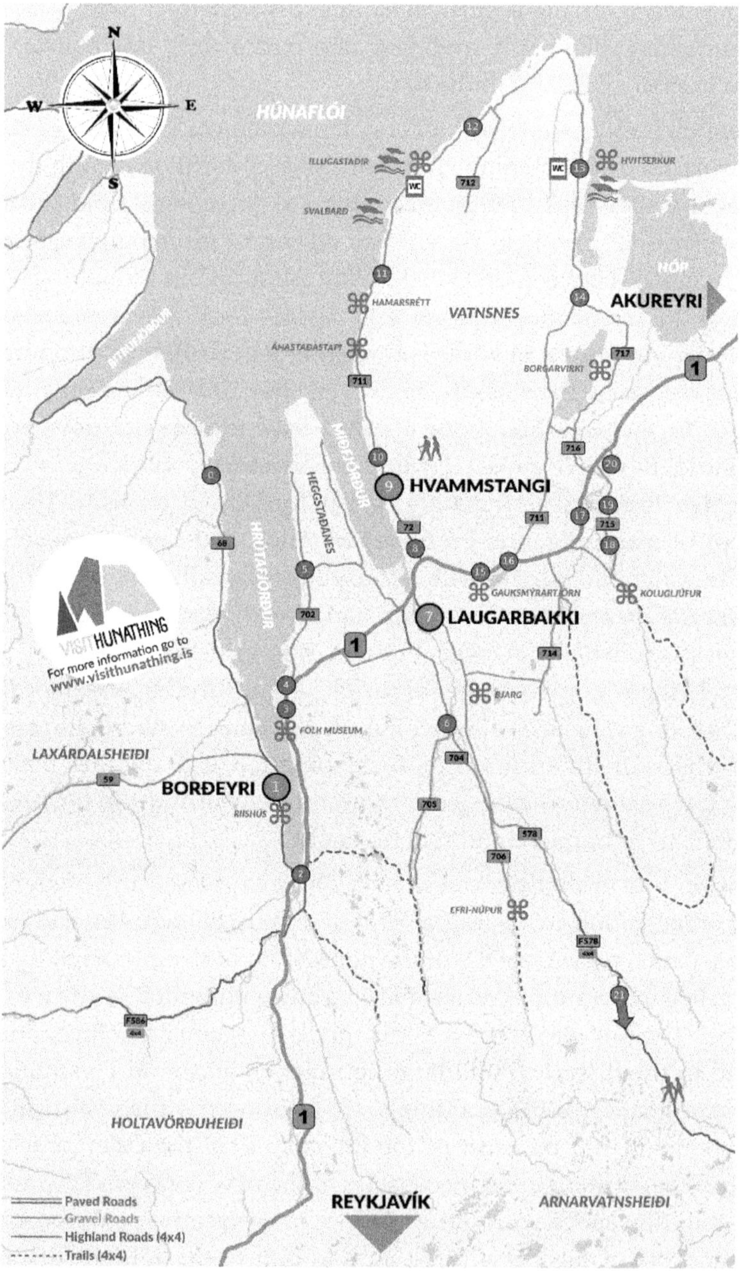

Figure 5.2: Visitor Map of Húnaþing vestra. Reproduced with permission from Visit Húnaþing.

Before the ICS was established, there was misinformation surrounding seals in contemporary times in Iceland, and research to further understand their behaviour was needed. For example, seals were thought to hurt salmon stocks, which led to seal culling at important Atlantic salmon estuaries, and may be a contributing factor in the decline of the Icelandic harbour seal population (Granquist & Hauksson, 2016). Another important collaborative research work conducted at the ISC with the Marine and Freshwater Research Institute is the annual seal count conducted throughout Iceland. Monitoring has shown that the harbour seal (*Phoca vitulina*) population has recently experienced a severe decline, the conservation status is critically endangered (Granquist & Hauksson, 2019a) and the grey seal (*Halichoerus grypus*) population is critically endangered (Granquist & Hauksson, 2019b). These examples further underline the importance of the collaborative work currently being conducted at both the local and national level for the protection of natural assets.

In 2007 and 2008, the ISC successfully lobbied the central government to create two specialist positions to run the ICS' research division: one to lead the seal research department in coordination with the Marine and Freshwater Research Institute; and the other to lead the tourism research department in coordination with Hólar University (Figure 5.3).

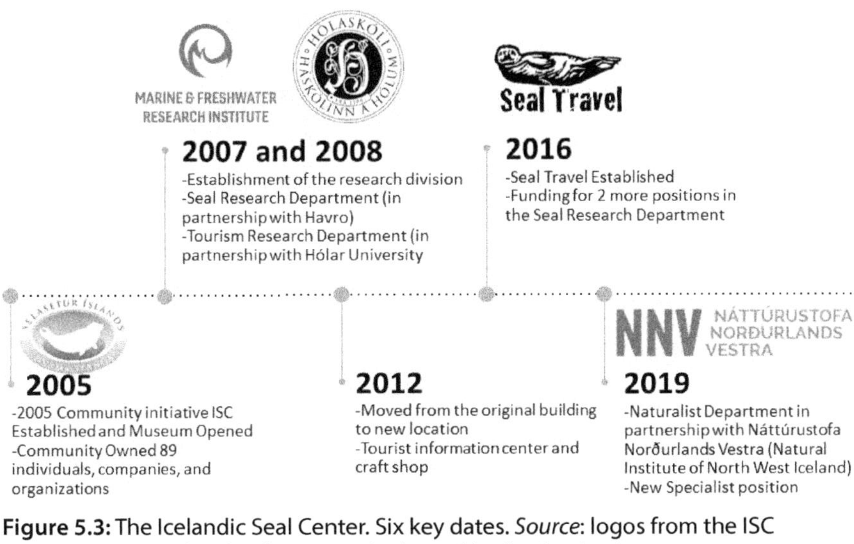

Figure 5.3: The Icelandic Seal Center. Six key dates. *Source*: logos from the ISC (https://www.northiceland.is/en) and adapted by Jessica Aquino.

Creating the two departments has had a positive impact on community understanding of seals, seal management and protection, and sustainable and responsible wildlife tourism development. Furthermore, job creation, including specialist positions, has further added to the positive impact on the community, because it began to build human capital while also helping to build community capacity.

In 2012, the ISC moved from its original building at Sigurður Pálmason to its current location near the harbour, where it is used as a tourist information centre and seal museum. In 2018, because of a growing staff, the research division was relocated to administrative offices near the centre of Hvammstangi. In 2019, the ISC had grown to six full-time positions and between one and three part-time, seasonal staff. The transformation of the old slaughterhouse and freezer into a museum and visitor information centre in 2012 began revitalising the harbour area. In March 2015, Sjávarborg Restaurant opened above the new museum. In 2016, to further develop tourism in the region, the ISC started a non-profit travel agency called Seal Travel dedicated to helping local entrepreneurs in Húnaþing vestra and North Iceland to market themselves nationally and internationally, and to create a network of tourism products with an emphasis on wildlife and seals. In 2019, a new naturalist department was created within the ICS research division, and a new specialist position was created in coordination with the Natural Institute of Northwest Iceland, *Náttúrustofa Norðurlands vestra,* with a focus on birds, specifically the eider duck, which is particularly valuable to farmers for its down feathers.

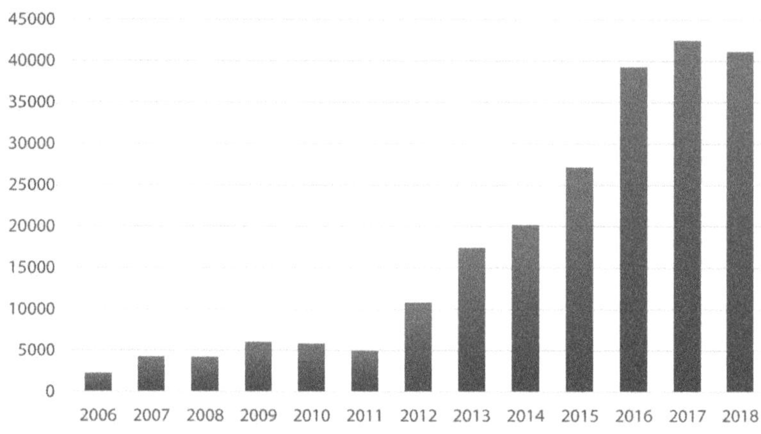

Figure 5.4: Visitor numbers to the Icelandic Seal Center from 2006-2018. Source: Reproduced with permission from The Icelandic Seal Center.

From 2006-2013, the ISC was open to the public during the summer season (May-September). However, as visitor numbers to Húnaþing vestra have continued to grow, in 2014 the ICS began opening from

February to December, and since 2015, has been open year-round. In 2011, when the ISC took on the responsibility of the region's Tourism Information Center, it received 4,958 visitors, while in 2012, visitor numbers to the ISC more than doubled welcoming 10,766 visitors (Figure 5.4). However, not all visitors enter the museum, as the majority use the information centre while, on average, only 35% of visitors enter the museum and pay admission (The Icelandic Seal Center, personal communication). In 2018, a total of 41,078 visitors were recorded with a 3% reduction in visitors from 2017. The majority of visitors come during peak summer season (May through September) with total average visitor numbers in July and August around 10,000 visitors each month.

With the creation of Seal Travel, and through collaboration of the research division and Tourism Information Center, the ISC aims to:

- Promote the development of tourism in the region through research;
- Collaborate with tour operators and entrepreneurs;
- Promote the transfer of knowledge between the scientific community and society;
- Inform tourists about responsible seal watching methods and viewing locations; and
- Develop environmental policy and seal management.

(The Icelandic Seal Center, personal communication)

The ISC continues to strengthen regional development through networking, capacity building, research, and innovation based on sense of place. The ISC is managed and staffed by residents and continues to cooperate with the community in its activities, which includes an annual seal count and public lectures. It also collaborates with landowners and other businesses in the operation of the seal-watching sites.

Similar to the GSP, the ISC has examples of neolocalism, such as community revitalisation through ecotourism and sustainable rural tourism development (Table 5.2). Through seals and other natural assets, the ISC has encouraged the development of a sense of place and local brand identity. For example, the Vatnsnes peninsula is known in Iceland, and internationally, as the 'the Land of Seals'. The ISC has also helped in creating permanent job opportunities in the area and has had a strong positive impact on the community in general. The seal museum has helped in the preservation of local history and local knowledge of seals.

Table 5.2: Neolocal elements of the Icelandic Seal Center

- ☐ Rural development through tourism
- ☐ Preservation of local history
- ☐ Community revitalisation through ecotourism and sustainable rural tourism
- ☐ development
- ☐ Community engagement and empowerment
- ☐ Sense of place and local identity
- ☐ Use and preservation of local knowledge
- ☐ Building community capacity

Discussion and conclusion

The community of Húnaþing vestra has used a renewed sense of place as a marketing strategy to increase revitalisation efforts and regional development. Both the GSP and the ISC were successful in attaching cultural and natural heritage to emphasise a connection to the distinctive locality of their region. In doing so, they were able to strengthen a sense of place while also preserving natural and cultural heritage. Locally made and grown products are produced and sold in Húnaþing vestra. However, in the tourism industry, these products have a greater reach, nationally and internationally. Both the GSP and the ISC, as community initiatives, have produced a sense of pride that comes from locally made products. Although the GSP is no longer a functioning project, it serves as a case study on how community engagement led to empowerment and local support. The ISC has a similar strategy, however early successes with lobbying the government and other funded research-based and community development projects helped the ISC to survive the financial crisis of 2008. The ISC continues the push for tourism regional development in the Northwest of Iceland through building community capacity, networking of tourism operators, and other related services.

Unintentionally, both the GSP and the ICS have used neolocalism to help revitalise a rural community. The GSP has had a significant impact on the local identity of Iceland's Northwest region. Both the GSP and ISC were implemented as a means for economic development. Both projects were used as a strategy to attract a greater number of tourists to Húnaþing vestra by using and strengthening existing community assets. By focusing on local knowledge, cultural heritage and traditions, and community assets, both community initiatives helped to mediate globalisation by focusing on the local. Shortridge defines neolocalism as, 'a deliberate seeking out of regional lore and local attachment by residents (new and old) as a delayed reaction to the destruction... of traditional bonds to community and family' (1996, p. 10). However, what was seen in our study was the deliberate seeking out of local assets that led to the preservation of natural and cultural heritage and helped to strengthen the sense of place and local identity of a community.

The use of community capitals led to the reversal of a 'spiral of decline' of a community that was threatened in North Iceland. A 'spiral of decline' is when there is a decline in one capital, which may trigger declines in other community capitals (Emery & Flora, 2006). Outmigration of young individuals and families highlight the challenges that small rural communities face in a changing economy. Since 2000, Húnaþing vestra has experienced a revitalisation of the local economy while more entrepreneurs have emerged using a sense of place and local identity as a strategy for development. The increase in visitation to Húnaþing vestra has both directly and indirectly affected services within the community (Burns, 2018). For example, during the summer season, the general store in town is open every day with longer hours and offers a larger selection of items. Þvottahús Perlan has grown from a one-woman cleaning service in 2005 to employing ten year-round staff with 15 summer season staff. There is a potential for food tourism as there are now five restaurants and two cafés (Hlaðan and Langafit) spread throughout the municipality, some of them open all year round.

There is also a growing arts scene, as the popular *Eldur í Húnaþing* festival is held each year in Húnaþing vestra, along with an increasing number of performing artists living in the municipality. Researching neolocalism and community festivals as a form of community revitalisation through culture and arts may add to the understanding of how these place-creating narratives are created, and may lead to further understanding of the process of community engagement and its connection to community empowerment, as shown in our two case studies.

References

Adger, W.N. (2000) Social and ecological resilience: Are they related?, *Progress in Human Geography*, **24**(3), 347–364.

Ahmad, M.S. & Abu Talib, N.B. (2016) Analysis of community empowerment on projects sustainability: moderating role of sense of community, *Social Indicators Research*, **129**(3), 1039–1056.

Aquino, J.F. & Burns, G.L. (in press) Creative tourism: The path to a resilient rural Icelandic community, in K. Scherf & G. Edwards (eds.), *Creative Tourism and Sustainable Development in Smaller Communities*, Calgary: University of Calgary Press.

Argent, N. (2018) Heading down to the local? Australian rural development and the evolving spatiality of the craft beer sector, *Journal of Rural Studies*, **61**, 84–99.

Björnsdóttir, R.E., Sigurbjörnsson, H.I., & Kristjánsdóttir, F. (2018) *Konan og SelshamurinnI* [The Woman and the Sealskin]. Kópavogur: Töfrahurð.

Burns, G.L. (2018) Searching for resilience: seal watching tourism as a resource for community development in Iceland, in A.A. Lew & J. Cheer (eds.), *Tourism Resilience and Adaptation to Environmental Change:Definitions and Frameworks, Routledge Advances in Tourism Series* (Advances i), Abingdon, UK: Routledge.

Byock, J.L. (2001) *Viking Age Iceland*, London: Penguin Books Ltd.

Chambers, C., Helgadóttir, G., & Carothers, C. (2017) 'Little kings': community, change and conflict in Icelandic fisheries, *Maritime Studies*, **16**(1), 10.

Conger, J.A. & Kanungo, R.N. (2011) The Empowerment Process: Integrating Theory and Practice, *Academy of Management Review*, **13**(3), 471–482.

Destination Viking Sagalands. (2005) *Destination Viking Sagalands: The Icelandic Sagas and Oral Tradition in the Nordic World*, in D. Cooper, R. Guðmundsson & T. Muir (eds.), Reykjavik: Destination Viking Sagalands, EU Project.

Eberts, D. (2014) Neolocalism and the branding and marketing of place by Canadian microbreweries, in M. Patterson & N. Hoalst-Pullen (eds.), *The Geography of Beer*, Dordrecht: Springer Netherlands, pp. 189-199.

Emery, M. & Flora, C. (2006) Spiraling-Up: Mapping community transformation with community capitals framework, *Community Development*, **37**(1), 19–35.

Flack, W. (1997) American microbreweries and neolocalism: 'Ale-ing' for a sense of place, *Journal of Cultural Geography*, **16**(2), 37–53.

Fletchall, A.M. (2016) Place-making through beer-drinking: a case study of Montana's craft breweries, *Geographical Review*, **106**(4), 539–566.

Freire, P. (1970) *Pedagogy of the Oppressed. Internal Control SE - 5*, New York, NY: Continuum.

Granquist, S.M. & Hauksson, E. (2016) Diet of harbour seals in a salmon estuary in North-West Iceland, *Icelandic Agricultural Sciences*, **29**, 7–19.

Granquist, S. M., & Hauksson, E. (2019a). Aerial census of the Icelandic grey seal (*Halichoerus grypus*) population in 2017: Pup production, population estimate, trends and current status, Marine and Freshwater Research Institution HV 2019-02. Reykjavik.

Granquist, S.M. & Hauksson, E. (2019b). Population estimate, trends and current status of the Icelandic harbour seal (*Phoca vitulina*) population in 2018. Marine and Freshwater Research Institution, HV 2019-36. Reykjavík 2019.

Holtkamp, C., Shelton, T., Daly, G., Hiner, C.C. & Hagelman III, R.R. (2016) Assessing neolocalism in microbreweries, *Papers in Applied Geography*, **2**(1), 66–78.

Icelandic Tourist Board. (n.d.) *Numbers of Foreign Visitors*. Available at https://www.ferdamalastofa.is/en/research-and-statistics/numbers-of-foreign-visitors. (Accessed: 19/07/2019).

Jóhannesson, G.T. & Huijbens, E.H. (2010) Tourism in times of crisis: Exploring the discourse of tourism development in Iceland, *Current Issues in Tourism*, **13**(5), 419–434.

Kokorsch, M. (2017) The tides they are a changin': Resources, regulation, and resilience in an Icelandic coastal community, *Journal of Rural and Community Development*, **12**(2/3), 59–73.

Moscardo, G. (2010) The shaping of tourist experience: The importance of stories and themes, in M. Morgan, P. Lugosi & J.R.B. Ritchie (eds.), *The Tourism and Leisure Experience: Consumer and Managerial Perspectives*, Bristol: Channel View Publications, pp. 43-58.

Moscardo, G. (2015) Stories of people and places: Interpretation, tourism and sustainability, in C.M. Hall, S. Gössling & D. Scott (eds.), *The Routledge Handbook of Tourism and Sustainability*, Abingdon, UK: Routledge, pp. 294-301.

Ólafsson, E. & Pálsson, B. (1805) *Travels in Iceland: Performed by order of his Danish majesty*, London: Richard Phillips.

Runolfsson, B.T. (1999) ITQs in Iceland: Their nature and performance, in *Individual Transferable Quotas in Theory and Practice*, in R. Arnason & H. Gissurarson (eds.), Reykjavik: University of Iceland Press, 103-140.

Schnell, S.M. (2010) Food with a farmer's face: Community-supported agriculture in the United States, *Geographical Review*, **97**(4), 550–564.

Schnell, S.M. (2013) Food miles, local eating, and community supported agriculture: Putting local food in its place, *Agriculture and Human Values*, **30**(4), 615–628.

Shortridge, J.R. (1996) Keeping tabs on Kansas: Reflections on regionally based field study, *Journal of Cultural Geography*, **16**(1), 5–16.

Statistics Iceland. (n.d.) Statistics Iceland. Available at http://statice.is/. (Accessed: 7/06/2019).

Stedman, R.C. (2003) Sense of place and forest science: toward a program of quantitative research, *Forest Science*, **49**(6), 822–829.

Tung, V.W.S. & Ritchie, J.R.B. (2011) Exploring the essence of memorable tourism experiences, *Annals of Tourism Research*, **38**(4), 1367–1386.

Valsson, T., Ulfarsson, G.F. & Gardarsson, S.M. (2013) A theory of the evolution of settlement structures based on identification and use of patterns: Iceland as a case study, *Futures*, **54**, 19–32.

Wallerstein, N. & Bernstein, E. (1994) Introduction to community empowerment, participatory education and health, *Health Education Quarterly*, **21**(2), 141–149.

Willson, M. (2016) *Survival on the Edge: Seawomen of Iceland*, Seattle, WA: University of Washington Press.

6 Community Festivals Reveal Tangible and Intangible Bounty: Lismore NSW Australia

Ros Derrett

Cultural life contains both the wealth of being able to appreciate and treasure traditions of all peoples and an opportunity to enable the creation and innovation of endogenous cultural forms (Agenda 21 2008: 8 cited in Perry et al., 2016: 23).

Abstract

Recent research suggests that social and cultural impacts, often understood in terms of community wellbeing and resilience, indicate that cultural festivals can stimulate job creation, tourism, attract visitors and investment and diversify the traditional focus on agribusinesses that holds many rural communities in decline. This chapter examines how a local volunteer run annual community festival, Lismore Friendship Festival's Piazza in the Park, encourages participation, reciprocity, and generosity amongst residents through a process of intergenerational engagement in a celebration of multiculturalism. If the cultural tourism is community-created, as is suggested by festivals welcoming visitors, the dynamic of working together to develop the community's image and identity may bind residents and encourage identification with community. In line with current literature, the impacts of this festival suggest that social capital is the product of social interactions with the potential to contribute to the social, civic or economic wellbeing of a community of purpose.

Introduction

The emerging literature on a sense of community, engagement, and empowerment serves this case study well. Recent research suggests that social and cultural practices, often understood in terms of community wellbeing, indicate that arts activity, including cultural festivals, stimulate job creation, tourism, attract visitors and investment, and diversify the traditional focus on agribusinesses that hold many rural communities in decline. When investigating regional culture and its reinvigoration, the phenomenon of neolocalism can manifest through the experience of groups celebrating their heritage in locations of a diasporic separate from their place of origin. Immigrant citizens can contribute to a host community, as it seeks to revive interest in preserving and promoting its identity.

My extensive personal and professional experience in festival-making research and practices has offered me opportunities to observe the strategic and tactical approaches groups have employed to conduct community festivals. My long-term interest in this phenomenon has prompted me to evaluate a festival, Lismore Friendship Festival, developed in my hometown in regional New South Wales (NSW), Australia. My determination to evaluate an emerging festival, initiated by residents to raise awareness of the contribution Italian settlers to the district in the 20th century have made to community life, is outlined in this chapter. The platform chosen to deliver this recognition was a festival. I have been an integral part of the team of volunteers who have now conducted the festival, with its roots in Italian communal life, for 5 years. It has afforded me a perspective informed by the practical and theoretical frameworks identified in this investigation of the nourishment festivals can provide for community resilience. These bear significant witness to the four main elements of sense of community suggested by McMillan and Chavis (1986: 14) of membership, influence, integration, and fulfilment of needs and shared emotional connection through sharing history and shared participation.

Positioning festivals and communities

Cohesion and quality of life, along with a sense of community and place, acknowledges belonging, valued membership, mutual importance and dependence, shared worldviews, bonding/networking, and mutual responsibility for the community as key indicators are demonstrated in this chapter's case study destination. These demonstrate vital dimensions, such as arts practice, sense of place, role of volunteerism, community resilience, and revitalisation of cultural heritage. Each of these elements provides a timely lens for an investigation into what is now regarded as neolocalism. Sarason (1974) proposed the existence of a psychological sense of community, asserting it is one of the major bases for self-definition, that people knew when they had it and when they didn't. This draws on the notions of interdependence, mutual responsibility, and collective consciousness. A renewed understanding of these frameworks has implications for community festival organisers attempting to connect with, and reflect, their host community to be shared with visitors. The festivals they frame can address the histories, heritage, and shared memories of particular host communities and attract visitors to acknowledge the relationships that informed the initial design and delivery of the special occasion. Central to most community festivals is an appreciation of a sense of place and the influence of its origins and meaning for residents and visitors, bringing it all back home!

Image 6.1: Lismore Festival graphic. www.LismoreFriendshipFestival.com.au

The Lismore Friendship Festival hosted in regional Australia, identified in this chapter's case study, focuses on belonging to Lismore and celebrating historic and contemporary life while encouraging the whole community to participate in diverse inclusive activities that can grow the community's capacity to create harmony. It aims to grow awareness and understanding of the contribution Italian settlers and their descendants have made to the city's quality of life. It aims to attract visitors to the city through a dynamic program, in conjunction with numerous city-wide stakeholders, to consolidate long term positive social, cultural, and economic impacts. It aims to make a valuable contribution to raising the city's profile as a destination that knows how to party!

I have argued (Derrett, 2007) that festivals allow people to reflect and determine a sense of community and place, represent their image and identity, and contribute to cultural tourism. This case study demonstrates that community-driven festivals celebrate the community's social identity, its historical continuity, and its cultural resilience. They are socially constructed and negotiated phenomena that can be staged in everyday places that also become tourist places. Festivals provide a forum for creativity, custom, heritage, and cultural practices for both resident and guest. I contend that festivals nourish community resilience by making inclusive celebrations for host communities and guests, unleashing relationships between many stakeholders through collaboration and co-operation. Each participant contributes to the program, traditions, cultural practices, impact, and reach of events with differing voices and emphases.

By confirming the local culture, a festival can embrace an adaptation that demonstrates a friendlier persona through embracing aspects of a distinctive subculture of the city by new settlers over the years. The interaction that has been experienced in the daily life of residents, regardless of their ethnic origins, becomes a relevant stimulus to simultaneously examine diversity, develop hybrids, and foster innovation that impacts the broader community. This blending of elements of different cultures serves to allow displaced ethnic groups to celebrate a heritage that is initially alien to their own origins. This reshaping is a common phenomenon as agencies develop inclusive policy frameworks to rediscover the notion of national identity through multiculturalism.

A sense of community relies on a feeling of involvement in experiences that generate networks of trust, like festivals. It makes the global feel local or 'glocal' (a buzzword of recent origin). This engagement can overcome the alienation and exclusion that can minimise the contribution each new settler makes to the diverse communities to which they belong through housing, schooling, workplaces, and recreational activity. The multiple identities come from choices each makes about where they may place the emphasis of active community engagement, and the degree of fluidity with which they voluntarily acquaint themselves. The festival interface offers individual benefits to develop new skills and personal recognition, through active participation or through leadership, by identifying and respecting established cultural practices while not compromising what their own cultural heritage has to offer. This allows for the creation of new narratives in the public sphere. The uniqueness of communities, like that of individuals, can be regarded as an accumulation of experiences.

> *Ubuntu – …speaks of the fact that my humanity is caught up and is inextricably bound up in yours. It speaks particularly about the fact that you can't exist as a human being in isolation. It speaks about our interconnectedness. You can't be human all by yourself. A person with Ubuntu is welcoming, hospitable, warm and generous, willing to share.* (*An age-old African ideal)*

A systematic approach to determining any enhancement of community spirit and heightened sense of place, or increased numbers of personal and societal networks within the community and beyond, or an upsurge in family connectivity can assist in assessing the lived experience of festival attendees. 'Social capital', said Eva Cox in 1995 (cited in Falk 1999: 113), 'should be the preeminent and most valued form of any capital, as it provides the basis on which we build a truly civil society'.

Lismore Friendship Festival organisers need to integrate personal, often ephemeral, experiences of the diaspora in which the new settlers now reside. The mainstream heritage conversation that links meaning to the heritage tourism paradigm can revive the cultural and social life of local people, build community pride, provide a sense of identity, increase community participation and pride of the area, or enhance the image of a destination as it becomes integrated in the tourism sector. The whole destination becomes visible as an event takes place encouraging visitors 'to do what the locals do'. The chapter examines how a local volunteer run community festival encourages participation, reciprocity, and generosity amongst residents through a process of intergenerational engagement in the celebration of multiculturalism.

Festival research – setting the scene

Getz (2007) reviews contemporary thinking on the social and cultural phenomena that are festivals. He summarises these as a sacred or profane time of celebration, marked by special observances; an annual celebration of a notable person or event or the harvest of an important product; a cultural event consisting of performances of works in the fine arts, often devoted to a single artist or genre; a fair; and generic gaiety, conviviality, cheerfulness. A defining characteristic of a festival is its *transience,* whether spontaneous or planned. Memories are built. Festivals commemorate foundation elements of particular social groups and become part of their cultural heritage. Festivals also link landscape to lifestyle in simple and complex ways where space becomes place. Falassi (1987) suggests that both the social function and symbolic meaning of the festival are closely related to a series of overt values that the community recognises as essential to its ideology and worldview, to its social identity, its historical continuity, and its physical survival, which is ultimately what festivals celebrate.

While MacCannell (1973) raised the notion of local culture being transformed into a commodity and becoming a staged culture, Huang and Stewart (1996: 30) suggested residents develop bonds as a result of the psychological investment of working together developing the town's image. If cultural tourism is community-created, as is suggested by festivals welcoming visitors, the dynamic of working together to develop the community's image and identity may bind residents and encourage identification with community. Jaeger and Mykletun (2013: 213) also concluded that festivals do influence the identities of people and place, and festival engagement influences the need to belong. They particularly highlighted how the connection between identity (of self, with others and wider society) builds a bridge between the personal and public worlds. Their research reinforced the notion that festivals can contribute to individual and collective identities, strengthen a sense of cohesiveness and belonging to place, while making these places more visible and increasing their capacity to attract people. Others, Karlsen (2007) and Snow (2001), also differentiated between personal identity, social identity, and collective identity as features accommodated during festivals.

A community comprises people who identify themselves as a group because of their shared cultural heritage, spirituality, geographic location, special interest, or gender (Peck 1990: 60). The Lismore Friendship

Festival focuses on a community defined as a group of people living near one another, in a social relationship in which intent, belief, heritage resources, preferences, needs, risks, and a number of other conditions are present and common. Communities trying to find common ground can utilise the festival's overt collaborative function to better plan for community resilience, as Miles (2008) and Manzo and Perkins (2006) have observed.

Community festivals can raise spirits in host communities by engendering civic pride as hospitality and tangible assets are shared by custodians of the community's capital with visitors. They can act as a catalyst for the support and promotion of strong, vibrant communities that highlight a distinctive sense of place for residents and visitors while celebrating cultural diversity. They stimulate distinctive place-making for festival-hosting destinations through purpose designed signage, ephemera, public art, responses to seasonality, and encouraging clusters within and between creative industries. Communities hosting festivals use them to feature the talents, experience, or skills of locals, or an opportunity for creative exchange by inviting others to join in through competitions, performances, displays, and free interaction.

Some scholars have argued that some community festivals, rather than encouraging diversity, can reproduce and disguise power relationships, leading not to increased tolerance, but rather to tensions and heightened differences. Organisers of the Lismore Friendship Festival have worked to minimise any disruption by highlighting the heritage values of the wider community while hosting themed activities to represent the long-standing connection between the city and the antecedent communities in Italy. The program addresses traditional and contemporary social and cultural themes, both in the host community and the overseas partners. This is driven by the aims and objectives of the auspicing organisation's strategic plan. Contributors to *Accessibility, Inclusion and Diversity in Critical Studies,* (Finkel et al., 2018) identify the challenges that are faced by organisers of multicultural festivals by examining the fragility, disruption, marginalisation, disharmony, and divergence that can emerge if care is not taken through inclusive purposeful planning and management.

Researchers Richards (2018), Sharpley and Stone (2011), Ui et al. (2016), and Wallstam et al. (2018) have responded to the growing evidence of important social impacts of community festivals as part of the local creative economy and resident quality of life. These issues recognise the depth of connection of such festivals to the regeneration of neolocalism.

The essence of community festivals today – what do community festival attendees want?

Festivals have emerged in response to a number of factors in contemporary society. There is recognition of an increased affluence of some sections of society that has led to increased resources available to events. The growth of the serious leisure concept has been appropriated across all ages and stages. The theoretical framework initiated by Stebbins (1982) helps explain the complexities of modern leisure as both an activity and an experience. It brings together concepts of serious leisure, casual leisure, and project-based leisure demonstrated by community festivals. It observes themes that emerge from festival consumption, gender relations, social capital, and the quality of life of amateurs, hobbyists, career volunteers, and occupational devotees.

As a result, audiences expect to have a high quality, authentic experience, and this has stimulated the market for entertainment and events in regional communities. For those concerned with greater rigour in the tourism mix, events have fulfilled an essential stimulus in destination marketing and management. An expanded corporate involvement in events at all levels that address the demographic and psychographic target segments enhances brand loyalty, even in small boutique community events, and has led to growth in corporate and community sponsorship. Policy and planning outputs through government at all levels has begun to recognise the community's response to hosting and participating in special-interest events. This is seen to result in increased awareness of the ability of events to attract tourists and extend their length of stay within a region, especially where festivals collaborate with one another to build an attractive 'package'. The ability of events to generate economic benefits and create jobs has increased government support for, and involvement in, events, as demonstrated by the formation of government-based event corporations in many jurisdictions; and increased media interest in, and coverage of, events (Derrett, 2016: 9).

These characteristics manifest themselves against a backdrop of lasting impacts of globalisation. There is a concern that communities and the places that sustain them have foregone the building blocks that served to create their uniqueness. In an attempt to reinvigorate the local, tell their own stories, and emphasise their own traditions, clusters of neolocalism are beginning to thrive. Festival organisers need to respond to attendees' desire for authentic experiences. The co-creation that is manifest with community festivals highlights the emergence of a new way of looking at heritage and generates renewed ownership of stories and memories that are fundamental to the notion of neolocalism. This focus on cultural leadership can be distributed across the community's social, economic, and environmental portfolios.

The individual's festival satisfaction is determined by such festival attributes as feeling validated, respected, and accepted. Research at the case study festival indicates that individuals and groups enjoy exposure to something creative, innovative, risky, stimulating, and memorable. They appreciate interaction with like-minded patrons and host community through a sense of sharing and belonging. A festival program that exposes respondents to heightened levels of fun, humour, mystery, wonder, spirituality, and magic can be shared on an intergenerational level, as well as independently, which is appealing. Family group responses that identify the feeling of safety and security, through efficient, effective, and aesthetically comfortable personal and crowd management, is key. There is recognition, too, of important opportunities to learn and appreciate new skills, receive value for money for access to specialists, and find spaces within the festival space for quiet and intimate rest and physical comfort. With digital technological advances, there is increasing connection through clear, compelling targeted marketing messages, while at the venue contact with friendly, knowledgeable, and accessible hospitable staff and volunteers is essential (Derrett, 2016: 11). Participants enjoy being the first in their social group to undertake an activity. They may use social media to share a festival well after the concrete interaction of their participation on site.

Crompton and McKay (1997) investigated visitors to festivals and attempted to identify the set of motives that stimulated visitors to attend festivals and to assess the extent the perceived relevance of motives changed across different types of events. Their framework for studying the motivation of festival visitors embraced five socio-psychological motivational domains:

- *Cultural exploration* – a desire to gain knowledge and expand intellectual horizons;
- *Novelty/regression* (seek) – a desire to seek out new and different experiences through pleasure motivated by a need to experience thrill, adventure and surprise; and a desire to engage in behaviour reminiscent of an adolescent or child;

- *Recover equilibrium* (escape) – a desire to refresh oneself mentally and physically from day to day stresses; and to alleviate boredom;
- *Known group socialisation* – a desire to interact with a group and its members, and to enhance family relationships and friendships, and a desire to strengthen or establish cultural identity; and
- *External interactions/socialisation* – a desire to be with, observe, and meet new like-minded people. These conditions consistently feature in the planning and management of community events.

Ancestral homelands continue to influence the identities and tourism behaviours of populations living in a diaspora. Festivals create a time and space to reflect on the collective myths where aspects of emotional memory can find creative expression through communal celebration with others (Mackellar & Derrett, 2015: 227).

Case study

Whose culture and whose heritage are celebrated at the Lismore Friendship Festival's Piazza in the Park?

This community festival acknowledges the identity of one group of participants and their degree of commitment to the community, as a whole, over time. In nearly every context, individual and collective behaviours have been found to balance between inclusion and exclusion; for the individual, a matter of choice; for the group, a matter of shared determination. Elements of the diasporic identities of Italian settlers have been re-imagined and reinvented through the Lismore Friendship Festival – Piazza in the Park and are shared by the event stakeholders. This festival's stakeholder community brings together organisers, entertainers, stallholders, young and old, residents and visitors, business and government, schools and community groups all wanting to demonstrate community pride in the cultural heritage of Italian settlers to the city. It is also connected to cities in northern Italy from whence the Italians settlers originated. The two cities are now sister cities through formal agreements with Lismore City Council, and residents of Conegliano and Vittorio Veneto are regularly updated on festival plans. Their traditional rituals, cultural observances, performances, and customs have informed the elements of the festival program. Annually, they receive invitations to participate in person or through online platforms to be shared at the festival.

These have become evident in the design, development, and delivery of the emerging annual Lismore Friendship Festival's Piazza in the Park, which demonstrates what Falk and Kirkpatrick (2000: 103-4) suggest are qualitative differences to the traditional locale-based notion of community. They propose a term 'communities-of-common-purpose', as community life is variable and defined by its purpose. They suggest that social capital is the product of social interactions with the potential to contribute to the social, civic, or economic wellbeing of a community of purpose.

The host community

The regional city of Lismore, situated on the Far North Coast of NSW, Australia, has demonstrated a great appetite for community celebrations over the years. The local district population of 46,000 people includes approximately 27,569 people within the city, with 83.3% born in Australia. The most common other countries of birth are England 2.0%, New Zealand 1.0%, Philippines 0.5%, and Italy 0.4%. The Lismore Local Government Area (LGA) has higher than average unemployment and has more socio-economically challenged populations when compared with state and national averages. This creates potentially vulnerable groups. Children and youth (0-19 years) comprise 24.3% of the populations, elderly (75+years) 7.9%, households where non-English language is spoken 6.7%, unemployed 3.6%, low income households 27.5%, culturally diverse 5%, internet not accessed from dwelling 6.7%, lone person households 11%, place of residence outside of LGA 5 years ago 24.1%, and born overseas 9.5% (Australian Bureau of Statistics, 2017).

A wide variety of services that support the growing population's needs in welfare, health, learning, social, biophysical environment, and creative pursuits have developed, and there is a willingness for residents to offer hospitality and security to newcomers. The city is home to a university, high schools, state of the art medical services, and buoyant creative industries, while attracting investment in infrastructure to meet business and recreational aspirations. Lismore is a vibrant and prosperous regional city.

In the past 5 years, the free, outdoor, family-friendly Piazza in the Park has set about gathering the local rural and urban populations to celebrate the heritage of Italian migrants that have made a lasting and deep contribution to this destination. Initially, many new settlers worked the land and helped build the now well-established tropical fruit industry.

While the city's settlers (1920s – 1970s) were a significant cohort of up to 400 families over time, they arrived after the establishment of a historic settlement 50kms to the south in the 1880s, established at New Italy by survivors of the ill-fated Marquis de Ray's expedition from northern Italy. Each wave of migration built the connections between their homelands and the new lands they embraced, initially through hardships, but ultimately to become recognised for their forbearance and resilience and their capacity to enliven the host communities. Commemorative signage at early sites now physically mark the recreational, cultural, and agricultural contributions made. Historic New Italy Museum (www.newitaly.org.au), the Italo Australian Sport and Recreation Club, and a mosaic highlighting the transit from the northern to southern hemispheres are part of the heritage manifest. In 1991, Lismore City Council formalised a friendship agreement with the cities of Conegliano and Vittorio Veneto, in the Veneto region, to acknowledge the links created by these migrants.

Distinctive features of case study festival

Lismore's Friendship Festival – Piazza in the Park is an annual celebration of the contribution Italians have made to the region. Each year, thousands of local and visiting attendees enjoy being Italian for a day with a twist of humour, fellowship, and active involvement in arts and leisure activities, along with conspicuous consumption of authentic Italian delicacies from Prosecco to pasta, pizza, cannoli, gelato, and good coffee! There are diverse arts and recreational activities in the lead up to the one-day Piazza in the Park, which is seen by organisers as providing tangible and intangible cultural capital.

Image 6.2: Lismore Festival welcome. Photo by Peter Derrett, July 2018.

The festival was initially established by volunteers working through the social and recreation Italo-Australian Club to prepare a free outdoor multicultural family-friendly event that included authentic Italian classic music, dance, food and wine, games (like bocce and tug-o-war competition), special children's activities, merchandise stalls, and exhibitions including Italian contemporary and vintage bikes. Based on outstanding success in terms of numbers, scale and depth of program design, the organisers have since established an independent incorporated association to address structure and strategy going forward.

The festival is a lively outdoor celebration in the city's central business district and is an integral part of a weekend packed with excitement that includes the iconic long-held Lismore Lantern Parade (www.lanternparade.com). The two free family-friendly events collaborate to attract visitors to the city. Both are dependent on the enthusiasm of the local community that has developed over years who share the resulting hospitality (Gursoy et al., 2004: 171). They each work hard to minimise overcrowding, as events become more popular, and mitigate the risk of authenticity being diminished as the process of increasing resident pride and cultural identity raises the city's profile.

The festival is an attempt to act as a catalyst to activate the city's quiet urban heart over a weekend. Through the celebratory mechanism, organisers have observed an enhanced community image while 'word of mouth' has played a significant role in encouraging community pride. The program has contributed a perceived consolidation of local culture and encouraged community cohesion through collaborating with community organisations committed to inclusion, community health and social wellbeing, education, and cultural development.

Image 6.3: Lismore Festival musician. Photo by Peter Derrett, July 2018.

Music as a potent symbol of Italian culture

The presentation of authentic traditional music in a public space is an effective linchpin for the festival's program. It has become the community's happy place. The communal dance encourages active participation to be shared across ages and stages. The outdoor parkland location replicates Italian public access and encourages new and old settlers to enjoin in a spirit of sharing and learning.

Festivals and events offer opportunities for communities to showcase and celebrate their culture, and yet the culture that is celebrated is not always endemic to the local socio-cultural environment. Festivals create a time and place for communities of interest to create their own sense of place and to openly share their social identity (Mackellar & Derrett, 2015: 230).

Image 6.4: Lismore Festival dancing. Photo by Peter Derrett, July 2018

Endemic cultural tourism is about the dynamic human environment that incorporates a spectrum of engagement with creation, custom, heritage, and activity and offers visitors opportunities to share a community's expression and experience especially through active participation in celebration.

Leveraging heritage and identity in tourism events

Perry et al. (2019) suggest that cultural heritage and intangible forms of cultural expression have been marginalised in contemporary urban development and absorbed into the economic discourse. They believe festivals provide an established platform to review their transience, ambiguity, changeability, and ephemerality sufficient to enrich our understanding of the relationships between hosts and guests. With increasing attention given to the nature, value, meaning, and characters of heritage (Ahmad, 2006), there is now room to evaluate how cultural identity is integrated into the discussion of sustainable tourism.

Tourism is grounded in culture(s) through its practices and manifestations. It exists independently for diverse reasons (Robinson & Picard, 2006: 17). Bouchenaki (UNESCO 2004: 17) suggests that 'without tourism, cultural diversity is not lived, experienced or felt through aesthetic emotions and a comprehension of values it contains and expresses'. Brida et al. (2012), Cohen (2008), Duffy et al. (2018), Fredline et al. (2003), Lill (2015), Jaeger (2019) all observe the value of cultural festivals in the broader community tourism offering and reflect on the social impacts on place and the tourist experience. The resources emerging from the social aspects of peoples' ways of life through beliefs, values, behaviours, and the environments they build that nourish culture and creativity. Tourists are drawn to the living context of cultures, as well as to the cultural heritage that underpins the diversity that cultural festivals promote. These resources are both material as well as symbolic and can be shared when the Italian theme is embraced by both residents and visitors.

The space, place, and time bound nature of community events can enhance their social, cultural, and economic impacts. The emerging relationship between concrete elements of community celebrations and their long-term implications of intergenerational memory, sense of belonging, and image and identity are absorbed by the broader host destination. The essence of the festival experience is readily translated into destination marketing plans for tourism entities. In contemporary regional Australia, there is increased attention given to social cohesion and cultural diversity, and the Lismore Friendship Festival offers opportunities to demonstrate and mediate the exchange that takes place at the interface of community celebrations.

The human resources that emerge from a community driven celebration energise and stimulate that sense of belonging. Arising from the Lismore Friendship Festival, for example, is such a research project, with significant long-term implications. The human dimension of commitment, curiosity, and life-long learning are unlocked, and inclusive local initiatives come into play. Individuals and special interest groups actively co-operate where elders, in this case early Italian settlers, will make a substantial contribution to the reinvention of the local narrative to ensure a greater understanding of their legacy. The project management will be undertaken by the Lismore Friendship Festival and feed into its future strategy by recognising the consequences for organisational planning and structure. It is anticipated that the collaboration will generate innovative and creative responses that will highlight memorable aspects of the Italian early settlement and reflect how this helped build the resilient future city. The resulting empowerment through face-to-face interactions can encourage greater advocacy from other ethnic minorities and more recent settlers to the region and grow the capacity of individuals and interest groups to establish the best fit for initiatives that represent the diverse mix. It aspires to represent what could called new localism.

Place changes people – people change places – women's experience of belonging to Lismore

This project addresses the growing awareness and understanding of the contribution migrants and their descendants have made to the city's quality of life. It celebrates the intergenerational experiences of settlers through a sense of place and spirit of community characteristic of resilience demonstrated by all residents. It examines the relationships between new and old residents and celebrates the impacts of the interaction over time. It seeks to illuminate the Local Government Area's (LGA) image and identity through the experiences of women.

- The starting point is the collection of personal reflections of real people who journeyed from Italy to Lismore to make a new life for themselves and their families and interpret a significant heritage of Italian settlement from 1920s to the present, their dislocation, arrival in the diaspora, adaptation, and their relationship with the land, the city and existing community.
- It is an inclusive, intergenerational, collaborative community building project focused primarily on women's experience. A selection of women will be interviewed and their stories will be interpreted through arts-led multi-media platforms for distribution within the community, regionally, nationally, and overseas.
- Initially, a cohort of volunteer interviewers will be trained to elicit the most significant personal stories of city residents that express the changes that may have influenced the sense of place and community that is now manifested in the individual and residents' quality of life.

- ☐ Specialist historians, visual artists, videographers and photographers will work with both the interviewers and interviewees to ensure a safe and effective sharing of knowledge and skills by gathering material that gives a snapshot of the LGA's social and cultural identity in the 21st century.

The project will:

- ☐ Produce interpretive, visual and audible narratives and storytelling from interviews with a range of women;
- ☐ Produce ethnographic maps of our villages and city areas of Italian and diverse migrant settlement;
- ☐ Assist a new generation of descendants to understand heritage through the voices of their families that can be shared with the wider community;
- ☐ Provide a digital and physical display and information to residents and tourists to our area on the rich and diverse 20th century history of settlement in Lismore;
- ☐ Promote and celebrate the diversity of the Lismore community through display materials at the Lismore Regional Gallery, Historic New Italy Museum, Lismore Regional Museum, Lismore Visitor Information Centre, and events such as the Lismore Friendship Festival;
- ☐ Collate content for a future reworked social history entitled The Power of Land – *Il Potere della Terra* (originally published in 1993); and
- ☐ Promote and strengthen Lismore's friendship relationship with Conegliano & Vittorio Veneto in Italy and other sister cities.

The project's success will be measured through:

- ☐ The improved understanding of the city's diverse cultural heritage and importance to old and new residents;
- ☐ Increased community participation in the research aspects of the collection of relevant stories;
- ☐ Enhanced content for future Lismore Friendship Festival programs;
- ☐ Documented history for future publications with the focus on 'Place changes people – people change places – women's experience of belonging to Lismore'; and
- ☐ Promotion of the final multi-faceted outputs to raise the profile of Lismore through links with museums, galleries, tourism, media, and education partners locally and internationally.

The clearly identified benefits for destination marketing and management will assist policy and planning by authorities. It will acknowledge how personal and professional local inputs (experience, wisdom, skills, and special interests) can deliver powerful 'bottom up' stories that can inform and empower participants. It can nurture lasting networks that can activate responses to community challenges. Such an example, in the same city, occurred in 2017 when the city needed to address a debilitating flood in its Central Business District (downtown precinct). It was evident during that period that the sense of community relied on a feeling of identification, maybe even multiple identities, where residents placed the emphasis on voluntarily being active within the community.

This involvement contributed to the alienation and exclusion some may have previously felt and helped build a network of trust. The shift from tangible to intangible engagement was manifested through skills sharing, information, and knowledge building, while the interactivity nurtured connectivity and translated into ongoing relationships between residents. Such collaborative activity is core to providing experiences that are 'slow adventures' that transform new values in regional communities.

Observations

> 'Festivals play a significant role in the life of a community. Where people have understood and used festivals to lift spirits, transfer knowledge and enhance neighbourliness, they have created community resilience. This means a robust community exists within the town, the community celebrates a bond and the town prospers. The celebrations that have been sustained through collaborative effort over long periods have seen the strengthening of communities. This can be observed from the levels of community activities, economic growth especially in terms of cultural tourism and the quality of life and wellbeing of residents. In short, the resilience observable through festival making demonstrates the cyclical nature of change that has been embraced without the destruction of the community' (Derrett, 2007: 6).

Community festivals that deal with heritage through engagement with the creative economy and the tourism sector increasingly embrace the tangible aspects of what was termed the 'experience' economy (Pine & Gilmore, 1998). While this approach is generally embedded in the tourism, leisure, and hospitality sectors, it did emphasise the value of educational, entertainment, aesthetic, and escapist approaches to product and service development. A distinctive element of festivals, such as the Lismore Friendship Festival, is the nature and role of volunteerism in its preparation and delivery. Individuals accrue valuable skills and experiences while building community capacity that can be applied more broadly.

Some link the growth of homegrown community festivals in the not-for-profit public domain with 'festivalisation' by situating them in the place-making and tourism sectors of the economy. It is argued that such events become tools for policy development, community renewal, repositioning, and destination branding from the bottom up. This overtly connects with the globalisation that volunteers on the ground seek to avoid by carving out a return to localised cultural production through shared community endeavours. In fact, many smaller social activities become aggregated to consolidate a commitment to local cultural and recreational practices that can be shared with visitors. This portfolio approach to programs appears to heighten the understanding of initial core objectives of festival organisers in building understanding and growing awareness of the host community's unique selling proposition. It positions the rediscovery of the past through contributing to the future identity of the host community, the essence of neolocalism. These cultural identities are strengthened by co-creation of festival content and help to deliver an authenticity that has growing appeal. So, while the festival is created for local audiences, the participatory nature of the program offers visitors an opportunity for direct engagement. Visitors along with local stakeholders, sponsors, benefactors, media, business enterprises, the education sector, regional alliances, and local government all become beneficiaries of the cultural leadership festivals can provide for the consolidation of the local image and identity based on their interpretation of neolocalism.

The experiential nature of festival-guest engagement builds personal testimony, and the increased access of social media means that traditional word of mouth assists in consolidating the position of the festival in the public domain. The positive image-making gleaned from the personal experience helps build local confidence amongst residents. This, of course, does not preclude the appeal of community events to commercial media, business stakeholders, and government investment if the links to policy can be clearly identified. Tourism enterprises, like hospitality, accommodation, transport, lifestyle entrepreneurs, arts, and infrastructure are all beneficiaries of any partnership with community festivals. There needs to be recognition that different actors involved in organising festivals have different objectives, which can lead to efforts being dissipated, or in extreme cases, to friction within the community (Mayfield & Crompton, 1995: 37). The cooperation that can be generated can be massaged to ensure residents demonstrate an increased willingness to support local business as a result of sponsorship or in-kind support observed at the time of the festival. This becomes a manifestation of a true celebration of community. What may begin as a social enterprise (something abstract or intangible), but becomes embraced as the embodiment of a sense of place.

The way forward

Community festivals can provide opportunities to build civic capacity while transitioning to a social identity that sustains the wellbeing of residents and builds community self-esteem, confidence, and pride. They can overtly demonstrate reconnection between old and new settlers, develop an openness to new knowledge of shared circumstances, and affect new relationships, both local and global, without diminishing the essence of their place and heritages. The intimacies that can occur within a community celebration can be of a scale that ensures individuals and groups make sense of their surroundings while planning for authentic responses to challenges that may confront internal or external stimuli.

When striving to thrive, the collective thinking requires thoughtful focus on life experiences through networks that may comprise family, groups, regional networks, corporate, and governmental sensitivity that is manifest in the design, development, and delivery of social enterprises like festival-making. The social change, often sought when global complexity is difficult to navigate on smaller scale destinations, is crucially tied to the resident's quality of life. Tourism is one dimension now intrinsically linked to the festival and event sector at a local level. It can raise the standard of living for residents, but it must not also undermine the key aspects of citizenship and belonging that are intangible benefits that nourish a host community's world view and restores aspects of this identity to make it culturally unique – emphatically the new localism.

References

Ahmad, Y. (2006) The scope and definitions of heritage: from tangible to intangible, *International Journal of Heritage Studies,* **12**(3), 292-300.

Australian Bureau of Statistics. (2016) *Quick Stat.* Available at https://quickstats.censusdata.abs.gov.au/census_services/getproduct/census/2016/quickstat/LGA14850?opendocument. (Accessed: 18/09/2018).

Bouchenaki M. (2004) Intervention de M. Bouchenaki Mounir, Sous Directeur général pour la culture at the Barcelona 2004 Universel Forum of Cultures. UNESCO, Paris.

Brida, J.G., Disegna, M. & Osti, L. (2012) Perceptions of authenticity of cultural events: A host-tourist analysis, *Tourism Culture & Communication,* **12**(2), 85-96

Cohen, R. (2008) *Global diasporas: An introduction,* London: Routledge.

Crompton, J.L. & McKay, S.L. (1997) Motives of visitors attending festival events, *Annals of Tourism Research,* **24**(2), 425-439.

Derrett, R. (2007) Regional festivals – Nourishing community resilience: the nature and role of cultural festivals in Northern Rivers NSW communities, PhD thesis, Southern Cross University, Lismore.

Derrett, R. (2016) *The Complete Guide to Creating Enduring Festivals,* Hoboken, NJ: John Wiley & Sons.

Duffy, M., Mair, J. & Waitt, G. (2018) Addressing community diversity, The role of the festival encounter, in R. Finkel, B. Sharp, & M. Sweeney (eds.), *Accessibility, Inclusion and Diversity in Critical Event Studies,* London: Routledge, 9-20.

Falassi, A. (1987) *Time out of Time: Essays on the festival,* Albuquerque: University of New Mexico Press.

Falk, I. (1999) *How do families connect with the wider society?* Community and Family Dynamics: The Role of Social Brokers in Social Wellbeing, Centre for Research and Learning in Regional and Rural Australia, p.3.

Falk, I. & Kilpatrick, S. (2000) What is social capital? A study of interaction in a rural community, *Sociologia Ruralis,* **40**(1), 87-110.

Finkel, R., Sharp, B. & Sweeney, M. (eds.) (2018) *Accessibility, inclusion and diversity in critical event studies,* London: Routledge.

Fredline, L., Jago, L. & Deery, M. (2003) The development of a generic scale to measure social impacts of events, *Event Management,* **8**(1), 23 – 37.

Getz, D. (2007) *Event Studies: Theory, research and policy for planned events,* Oxford: Butterworth-Heinemann.

Gursoy, D., Kim, K. & Uysal, M. (2004) Perceived impacts of festivals and special events by organisers: An extension and validation, *Tourism Management,* **25**(2), 171 – 181.

Huang, Y-H. & Stewart, W.P. (1996) Rural tourism development: Shifting basis of community solidarity, *Journal of Travel Research,* **34**(4), 26-31.

Jaeger, K. (2019) *Tourists and communities in Rural Festival Encounters, A mutually beneficial relationship?* PhD thesis, University of Stavanger, Norway.

Jaeger, K. & Mykletun, R.J. (2013) Festivals, identities, and belonging, *Event Management*, **17**(3), 213-226.

Karlsen, S. (2007) *The music festival as an arena for learning: Fetspel I Pite Älvdal and matters of identity*, PhD dissertation, Luleå University of Technology, Luleå.

Lill, G. (2015) *The impacts of events on local community*, Masters Thesis, Aalborg University, Denmark.

MacCannell, D. (1973) Staged authenticity: Arrangements of social space in tourist settings, *American Journal of Sociology*, **79**(3), 589-603.

Mackellar, J. & Derrett, R. (2015) Standing stones, festival and the Celtic diaspora, *Tourism, Culture & Communication*, **15**(3), 227 – 236.

Manzo, L. & Perkins, D. (2006) Finding common ground: the importance of place-attachment to community participation and planning, *Journal of Planning Literature*, **20**(4), 335-350.

Mayfield, T.L. & Crompton, J.L. (1995) Development of an instrument for identifying community reasons for staging a festival, *Journal of Travel Research*, **33**(3), 37 – 44.

McMillan, D.W. & Chavis, D.M. (1986) Sense of community: a definition and theory, *Journal of Community Psychology*, **14**(1), 6-23.

Miles, R. (2008) Rural-regional sustainability through revitalising the commons: A case study, paper presented at AARE Annual Conference, 28 June – July 1, Brisbane.

Peck, S.M. (1990) *The Different Drum: Community Making and Peace*, London: Arrow Books.

Perry, B., Ager, L. & Sitas, R. (2016) *Festivals as integrative sites, valuing tangible and intangible heritage for sustainable development.* Available at https://www.mistraurbanfutures.org/sites/mistraurbanfutures.org/files/festivals-as-integrative-sites-report.pdf. (Accessed: 15 /11/2018).

Perry B., Ager, L. & Sitas, R. (2019) Cultural heritage entanglements: festivals as integrative sites of sustainable urban development, *International Journal of Heritage Studies*, 1-16.

Pine, J. & Gilmore, J.H. (1999) *The Experience Economy: Work is Theatre & Every Business a Stage*, Boston, MA: Harvard Business School Press.

Richards, G. (2018) The creative economy, Entertainment and performance, in C. Cooper, S. Volo, W.C. Gartner, & N. Scott (eds.), *SAGE Handbook of Tourism Management: Applications of Theories and Concepts to Tourism, Volume 2*, pp. 315 - 327.

Robinson, M. & Picard, D. (2006) *Tourism, Culture and Sustainable Development*, UNESCO, Paris France. Available at https://unesdoc.unesco.org/ark:/48223/pf0000147578. (Accessed: 2/08/2018).

Sarason, S.B. (1974) *The Psychological Sense of Community: Prospects for a community psychology*, San Francisco, CA: Jossey-Bass.

Sharpley, R. & Stone, P.R. (2011) Socio-cultural impacts of events: meanings, authorised transgression and social capital, in S.J. Page & J. Connell, J. (eds.) *The Routledge Handbook of Events*, London: Routledge, pp. 345-359.

Snow, D. (2001) Collective identity and expressive forms, in N.J. Smelser & P.B. Baltes (eds.), *International Encyclopaedia of Social and Behavioural Sciences*, London: Elsevier Science, pp. 2212-2219.

Stebbins, R.A. (1982) Serious leisure: A conceptual statement, *Pacific Sociological Review*, **25**(2), 251-72.

Ui, S., Day, J. & Cai, L. (2016) *The influence of festivals and local events on community quality of life, Tourism Travel and Research Association: Advancing Research Globally*, 9. Available at http://scholarworks.umass.edu/ttra/2013/Acadamic-Papers_Visual/9. (Accessed: 28/08/2018).

Wallstam, M., Ioannides, D. & Pettersson, R. (2018) Evaluating the social impacts of events: in search of unified indicators for effective policy making, *Journal of Policy Research in Tourism, Leisure and Events*, 1-20.

Websites

https://lismorefriendshipfestival.com.au/ - Lismore Friendship Festival

http://www.newitaly.org.au/ - Historic New Italy Museum, Pacific Highway, Woodburn, NSW, Australia

http://www.lanternparade.com – Lismore Lantern Parade produced by LightnUp Inc.

7 Migration and Tourism Mobilities:
How Younger Koreans are Enhancing New Urban Experiences in São Paulo

Eanne P. Leite and Thiago Allis

Abstract

Bom Retiro, in Sao Paulo, Brazil, is identified as a multicultural zone. In the last few years, Korean residents in Bom Retiro have been promoting their cultural activities and updating their businesses. However, Bom Retiro has other tourism products linked to the Korean community, such as a distinctive shopping area. The goal of this research is to understand the ways in which the Korean legacy participates in tourism development, from the perspective of Korean bakeries. Data reveals that the Korean cultural legacy takes part in Bom Retiro tourism development by articulation of ethnicity from Korean entrepreneurs, and by the global process of cultural commoditization, thus producing consumption spaces at the local level, such as bakeries, gastronomy, fashion, music, and language'. and consequently attracting a broad diverse set of visitors interested in Korean culture.

Introduction

It is common sense to assume that the acceleration of globalisation leads to homogenisation of cultures, reshaping lifestyles and landscapes (Gotham, 2005; Harvey, 1989). However, by emphasising this discussion from another angle, the emergence of neolocalism might represent resistance movements and attempts at differentiation (Collins, 2007). In this context, if, on the one hand, the growth of migration depicts undesirable contingents (such as wars, poverty, religious intolerance), on the other hand, (re)settlement of ethnic groups and their cultural manifestations (way of life, music, gastronomy, language) elsewhere provide resources to transform given portions of major urban concentrations into spaces of singular expressions of 'ethnicity' (Timothy, 2002). From this point of view, many tourist destinations take advantage of such ethnic resources as inputs to develop activities capable of attracting tourists and visitors to particular precincts within the cities. This chapter aims to present and discuss how cultural manifestations of Korean immigration to São Paulo, Brazil intertwine with urban tourism in a global destination influenced by significant migratory flows since the 19th century.

Korean migration dates back 60 years in São Paulo, concentrating their economic, social, and cultural activities within the district of Bom Retiro, a geographic target of several migratory groups throughout the 20th century, including Jewish immigrants and, more recently, Bolivians. The traits of the Korean culture are explicit in the urban landscape, observed in typical restaurants, *Hangul* (Korean alphabet), markets, and churches. One of the reasons for this resistance to assimilation could be attributed to the concept of the 'ethnic enclave' (Chi, 2016), when locals' business and social life rely on the community itself, preserving symbols and signals of the Korean culture.

Today, younger Brazilian-Koreans, usually the second generation after the arrival of their grandparents, are undertaking vigorous commercial renewal, recognising the importance of cultural activities as assets for attracting (new) visitors. This new wave of cultural valorisation could be framed in the wider scope of neolocalism. If these commercial strategies enhance tourism and attract resources for urban development (Fainstein & Gladstone, 1999), it raises concerns about the processes of cultural commodification (Rath, 2007; Williams & Hall, 2000), involving both tourists and local visitors. Thus, this chapter will offer the opportunity to explore economic, urban, and cultural aspects of contemporary migratory phenomena through the lens of tourism, particularly by observing 'tourism mobilities' within a large metropolis of the Global South.

Background

São Paulo a global multicultural hub

São Paulo is the eighth most visited city in Latin America (Global Destination Cities Index, 2018), being a cosmopolitan and global hub 'of great tourist attractiveness due the diverse cultural and economic thrills' (PLATUM 2015-208: 16). Over the last 20 years, tourism has become an emerging activity, aligned with the economic diversification of large global cities.

As the financial and economical hub of Brazil, and in several aspects of Latin America, São Paulo has faced many changes at a variety of levels. With the development of new technologies and the turbulent global economic changes that took place during the second half of the 20th century, the competition between global cities (Carvalho, 2011) stimulated the development of new urban cultures as a response to the rising mobility of the creative classes (Florida, 2011). In this context, and over the last thirty years, these global urban centres have become more connected to the world economy. In turn, the importance of tourism has become more relevant to the economy of São Paulo.

Similar to many other global cities (Sassen, 1991), São Paulo's cultural production has evolved in order to attract and entertain incoming cultural elites, improving its prestige within the international arena (Cordeiro, 2002). However, it still faces drawbacks in its intentions to expand the tourism sector, mainly because this activity appears sporadically in urban policies, without proper planning for the sector. Moreover, tourism in São Paulo imposes itself as a spontaneous consequence of visitation to an attraction, detached from the need for large urban projects that traditionally focus on attracting international tourist flows (Allis, 2012).

Until the 1990s, the business segment was seen as the primary motivator in attracting tourists. However, following the consolidation of policies granted by the Federal Constitution (Brasil, 1988), cultural heritage (either material or immaterial) moved to the forefront of urban and tourism planning (Fonseca, 2009). This new focus on urban and cultural heritage generated a new perspective towards developing tourism to build a competitive, international touristic destination (Allis, 2012).

In this sense, the cultural complexity of this metropolis of 11 million people, which has received various nationalities of immigrants since the 19th century, has resulted in some areas of the city being shaped as particular territories of well-defined ethnic groups. Liberdade and Bexiga, for instance, are both emblematic given the migrant concentration, where urban changes have been influenced by policies of tourism promotion based on their cultural aspects through the making of ethnic experiences (Bitelli & Bastos, 2018; Bocci, 2009). More recently, the community of Bom Retiro is also following a trend of tourism development based on the idea of an 'ethnic neighbourhood', focusing on multiculturalism from several nationalities, specifically Korean immigrants as the more expressive group.

Bom Retiro: Where global and local blend

During the second half of 19th century, the railway between Santos and the inland cities of Jundiai made Bom Retiro and its surroundings a strategic point for the circulation of goods and people. Portuguese and Italians were the first immigrants to settle in the region, where the latter composed approximately 70% of the workforce for the new factories.

The Portuguese immigrants exceled in commerce, locating their warehouses at Immigrants Street (Dertônio, 1971), today Jose Paulino Street, which is still an industrial hub for the clothing sector. Additionally, Jewish migration increased in São Paulo, particularly after World War I. The first settlements were in Pará, Rio de Janeiro and Southern Brazil and were associated with commercial and industrial development. Jewish immigrants contributed to the growth of commercial activities in the district, establishing a distinguished character separate from the other working classes in São Paulo's districts at that time (Póvoa, 2007; Truzzi, 2001).

For many years, Jewish immigrants were part of the district landscape. It was common to see their ethnic clothing when walking in the streets, but also their commercial shops, such as clothing and grocery stores, which are still present today, although to a lesser extent. From the 1960s onwards, a new migratory flux began to change the district landscape. As a result of the Treaty on Basic Relations in 1963, a mutual agreement between the Brazilian and South Korean governments, Korean migration to Brazil increased significantly.

As part of the contract between the governments of Brazil and Korea, the newcomers would ideally be settled in the countryside of São Paulo State. However, Koreans did not adapt themselves to farming since most of them were former military or small business owners from the urban middle and upper classes, possessing a certain degree of education (Choi, 1996). In the 1970s, they migrated to the city of São Paulo and initially settled in the Glicerio area, near the centre of the city where many families opened small shops, spotlighting clothing making (Chi, 2016).

Image 7.1: Traditional Jews and Koreans strolling in Bom Retiro.
Source: http://www.saopauloantiga.com.br/sinagoga-kehilat-israel/.

Soon, Bom Retiro became attractive to small businesses, with its easy access to both the downtown consumer market and the available workforce (Truzzi & Sacomano Neto, 2007). These factors led to a significant transformation, particularly noticeable around Jose Paulino Street where there was a concentration of popular clothing shops and textile production, modernised mainly by Korean entrepreneurship, who imported raw material and new technologies from South Korea to support the increasing demand. As immigrants began to gain economic status, by assuming their own dressmaking facilities, new immigrants were attracted to the area, specifically Bolivians (Chi, 2016).

This succession of ethnic migration (Truzzi, 2001) historically marked Bom Retiro by its diversity and multiculturalism, which is still present today throughout the urban landscape, as seen in restaurants, markets, buildings, and conversations on the street. The Jewish community still owns the most real estate in Bom Retiro, whereas the Koreans are active in the clothing businesses (da Cruz Paiva, 2011). The Bolivians, hired as a workforce for the dressmaking ateliers (Freitas & Baeninger, 2010), are discrete throughout the district, almost invisible to the visitors' eyes.

Nevertheless, Koreans are primarily responsible for making this area an ethnic enclave (Chi, 2016), reinforcing a sense of territorialisation by protecting cultural, social, political, and economic development (Cymbalista, 2007). The Korean identity and cultural attributes are at the forefront of commercial enterprises, often seen in Hangul (the Korean alphabet) signage at the events and festivals which take place in the district, and at the monument honouring the community, inaugurated in 2018. In addition, the media increasingly suggests walking tours of Bom Retiro, highlighting cultural and gastronomic diversity in the neighbourhood as an attraction for both tourists and city visitors (Menezes, 2019).

Image 7.2: Monument in honor of Korean immigration URI. Photo by Eanne Leite, September 2018.

Korean ethnic legacy: From kimchi to bakeries

The estimated number of Koreans in Brazil is 50,523, including second and third generations, the undocumented, the temporaries, and the ones naturalised as Brazilian citizens (Antônio & Araújo, 2019). It is the seventh largest foreign community in São Paulo, with 16,979 persons officially registered (São Paulo Cosmópolis, 2017). Until recently, Korean entrepreneurs targeted their businesses mainly to their own community, whose members either work or live in the area. The Korean migrants and their descendants, who opted for not working in the textile sector, became either self-employed or business owners, supported by their own ethnic peers. For this reason, all types of services are offered in the community, creating a small-scale Korean society (Yang, 2011).

One can see the Korean influence in commercial enterprises, such as private language schools, grocery stores, beauty parlours, and restaurants with Korean language menus. These practices have contributed not only to the formation of ethnic enclaves, but also as a centre of the ethnic economy with spatial concentrations of commerce and services designated to support the local community's demands. Roughly, 75% of Koreans run their own businesses, and of these, 80% are linked directly or indirectly to clothing shops.

Image 7.3: Typical Korean restaurant in Bom Retiro. Photo by Eanne Leite, July 2018.

The presence of Koreans has changed the district landscape through its ethnic legacy. The grocery stores are only outnumbered by churches (Catholic, Protestant, and Buddhist). There are beauty parlours, electronics shops, flower shops, kindergartens, karaoke bars, thirty-two restaurants, five markets, an ice cream shop, and a Tae Kwon Do school, among others (see Image 7.4, Zone C). Cafes and bakeries are becoming meeting places for members of the Korean community and for the youth interested in learning about Korean culture.

For the locality, it is a way to keep its culture alive on a daily basis (Baldissera et al., 2017). This legacy, constituted by arts, dancing, clothing, written languages, food, and music, are reified into urban spaces, in cultural centres (where one can find the *hanbok*, typical Korean garb), cafes, bakeries, groceries, and restaurants, which have gradually become attractive to visitors and tourists who stroll through the district.

However, it is only recently that the daily routine of Bom Retiro has become a factor for tourism visitation. Over the last few decades (1970s-2000s), two linked processes have contributed to the consolidation of the outskirts of Bom Retiro as an attractive area for thousands of visitors: the heritagisation of Luz Station quarter (Zone A) and the emergence of José Paulino Street as an open-air shopping mall (Zone B). Additional cultural developments have been implemented since the 1970s in an effort to revitalise the downtown city area through 'cultural anchors' or cultural activities (Kara-José, 2007: 205).

Image 7.4: Cultural heritage (Zone A), shopping area (Zone B) and ethnic enclave (Zone C) at Bom Retiro. *Source*: adapted by Eanne Leite from data collected on field-work, displayed over GoogleEarth image, May 2019.

The Luz Station, a building inspired by Big Ben in London, was the main station of the former São Paulo Railway installed in 1867 to connect the coast to the São Paulo countryside. Currently, it is a hub of the subway and the metropolitan trains. Within the station complex is the Museum of Portuguese Language, which burned in 2015 and is in the final stages of restoration. On the weekends, many tours depart from Luz Station to localities within the metropolitan areas of São Paulo (Paranapiacaba, Mogi das Cruzes and Jundiai). In 2018, roughly 13,000 tourists were served (CPTM, 2018) by the Touristic Express.

The main art gallery of the State of São Paulo, called Pinacoteca, is housed in the 1896 building designed by the architect Ramos de Azevedo, which is now one of the main tourist attractions of the city, hosting permanent and temporary exhibitions with an international reach. In 2017, Pinacoteca received roughly 880,000 visitors (Pinacoteca, 2018). In 2015, the exhibition from the Australian Ron Mueck alone received 402,000 visitors (Genestreti, 2015). The Luz Park, close to Pinacoteca, was refurbished in the 2000s, with funds provided by the Inter-American Development Bank, strengthening tourism development policies of the urban city patrimony. In the mid-1990s, the São Paulo Opera Hall was housed in the Julio Prestes station, which was the main station of the Sorocabana railway.

These and other cultural spaces make up a complex set of attractions for cultural tourists, mainly based on the heritage of monumental characteristics, generally anchored in erudite cultural manifestations, like paintings, sculptures, classic music, and sacred art, among others. Such initiatives are expected to create conditions to leverage tourist attractions, 'not only for the locals as a whole, but also as a complementation to business tourism' (Vargas, 2000: 85). From another perspective, the Bom Retiro district houses the main artery of clothing shops and ateliers. There are roughly 2,000 shops, where 70,000 people circulate daily (Chamber of Shopkeepers, 2017).

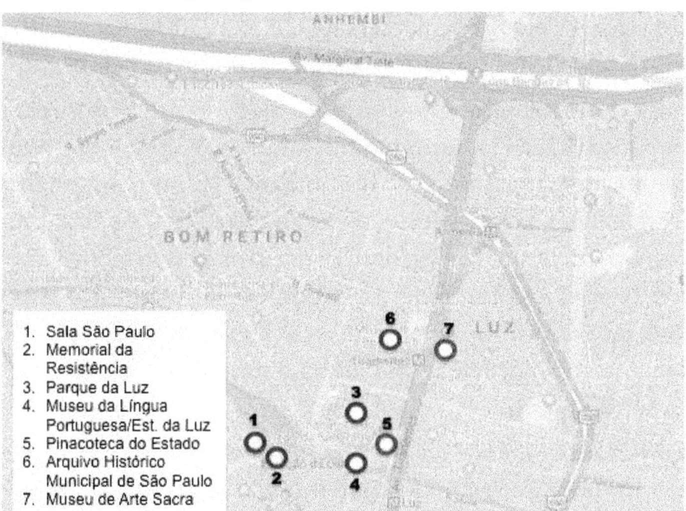

Figure 7.5: Zone A – Heritage venues at Luz Station quarter. Source: adapted by the authors from Google Maps October 2019.

The region is promoted as one of São Paulo's main shopping centres oriented towards consumers and other merchants from various parts of the country. For several years, the local tourism authority has been promoting shopping tourism in the city, with an emphasis on Bom Retiro. In this context, a new issue is the daily life of the community as an attraction. With the intense flux of people who visit the cultural centres (Zone A) and the shopping area (Zone B), the area with calm streets between shops and residences on the inner side of the district (Zone C) emerges as a potential destination. Many Korean entrepreneurs purposely rely on the symbolic and ethnic iconography of the public spaces (Collins, 2007), recognising its cultural legacy as a source of attractiveness.

Methodological approach

Between June and November 2019, Korean entrepreneurs and visitors at businesses considered 'ethnic' in Bom Retiro were interviewed (Gotham, 2002; Hall & Rath 2007). Recent renovations in Bom Retiro were designed to bring Korean ethnicity into modern shops and encourage Brazilian patronage. The interviews aimed to identify how the Korean ethnic legacy contributes to tourism development in relation to its material heritage and shopping experience. This approach helped to identify if the visitors' interests are primarily in the area of traditional heritage (well established in Zone A), shopping opportunities (Zone B) and to what extent these would be linked to the Korean legacy (with greater visibility in Zone C and the empirical focus of this research).

The interviews were conducted with two different groups: Korean entrepreneurs, such as restaurant and coffeeshop owners in Bom Retiro (primarily located in Zone C – see Image 7.4); and Brazilian visitors just after their visit to the selected Korean bakeries. The first group was chosen from a list of the Korean enterprises located in Zone C, while the other group was chosen in order to determine if the non-Korean visitors had knowledge about the area's cultural legacy, in light of the lack of advertisements or marketing materials about Bom Retiro as a tourist attraction of São Paulo.

As the literature is relatively silent on the context and topic of this research, with Bom Retiro as the case study, it was determined a qualitative approach would provide an advantageous format. When arguing in favour of qualitative research in social sciences, Lune & Berg (2017) state that:

qualitative research (...) refers to the meanings, concepts, definitions, characteristics, metaphors, symbols, and descriptions of things. In contrast, quantitative research refers to counts and measures of things, the extents and distributions of our subject matter: how large a thing is, how many of them there are, or how likely we are to encounter one (p. 12).

As this research is exploratory in nature, it was deemed a perspective where social phenomena are studied with minimal a priori presumptions of explanation would be appropriate (Lincoln & Guba, 1985). While in future research, an approach that combines both qualitative and quantitative could yield additional insights, the focus here was not to measure the phenomena but to properly identify narratives and descriptions, for example, in order to lay the groundwork for future studies.

Korean entrepreneurs were selected according two criteria: younger Koreans who speak the Portuguese language fluently, and those who renovated their establishments recently. For non-Korean respondents, the choice was made randomly as they visited Korean bakeries. In total, 26 people were interviewed among Korean entrepreneurs and non-Korean visitors. The questions were formulated in order to: observe the changes that have occurred in Bom Retiro over the past three years, such as when the former mayor of São Paulo proposed renaming the neighborhood to *Little Seul* (Venceslau, 2017); and determine if a non-Korean public frequents the bakeries. The interviews were composed of predetermined, standard questions. Responses were recorded and transcribed afterwards, highlighting the visitors' opinions and feelings about Bom Retiro. The contents of the interviews were directly interpreted by applying a basic content analysis, which was used to determine similarities, patterns, and significance. As an additional resource, basic cartographic tools (free versions of GoogleEarth and Coreldraw) were employed in order to highlight and study different spatial portions of the neighbourhood and identify how different 'spatial behaviours' (Edwards et al., 2008; Maitland, 2010; Selby, 2004) are spread throughout the urban fabric by these newcomers. Standard methods were employed to enhance the rigor of the data generation and the verisimilitude of the findings.

On the crossroads: how to keep identities and improve visitation?

Most of the entrepreneurs realise that the mix of people who visit their stores is changing: in the recent past, they were almost exclusively Koreans and their descendants. Now a broader audience is much more common. In response, some managers started to adapt to the new audience by gradually making small changes, like refurbishing their stores, restaurants, and coffee shops to reflect the new urban space. Bom Retiro is becoming more modern. These changes are small, and there is some reluctance on the part of the Korean business community to change too rapidly. For instance, the food sector has had to shape Korean food recipes to suit Brazilian tastes. Only one bakery owner has made significant changes to his bread and cake recipes to accommodate these new visitors, and other businesses have made small changes because they are struggling to import certain ingredients from South Korea.

There are also the innovators who are influencing Bom Retiro: youthful managers who bring in new Korean products, often driven by the K-Pop music scene, to the area in order to attract a specific audience. Many of these products have been noticed by social media. For example, an ice cream shop, located in the main shopping area, was the first franchise to bring the 'snow ice cream recipe', an innovative Asian style from Korea. Another example is a teashop, owned by a Brazilian who lived in Asia, that offers hot and cold Asian tea recipes and uses Portuguese and Korean menus. According to the interviews, both of these stores have a constant need to reinforce the ethnic features of the products as authentically Korean, primarily through their merchandising on social media. This supports product identification by consumers, as well as other ethnic elements in the area, highlighting an advantage of social media: it allows business owners to connect and communicate with their public.

As many locations and regions search for their own identity to protect themselves from the homogenisation caused by globalisation (Bell & Valentine, 1997), the Koreans, at least the majority of them, are trying to preserve their culture by including original recipes, maintaining traditional restaurants, and using their writing.

However, two subjects are forcing Korean entrepreneurs to evaluate small changes in their businesses in order to attract more visitors. First, as Korean families are the primary consumers of Korean establishments in Bom Retiro, the fall in the purchasing power of Korean families, mainly due to the recent Brazilian economic crisis, has focused attention on a broader consumer base. Second, Korean owners of newer bakeries are trying to reimagine their shops by seeking a wider audience and have included Portuguese signage and menus. Although influenced by western aesthetics, the set of ethnic elements, which is included in the food scene, are essential for the promotion of the immigrant culture and as a motivator for tourists (Plummer et al., 2005).

Interviews with visitors showed that some had previously heard of Bom Retiro and knew that it had a concentration of Koreans. This group, mostly young people, looked for groceries, coffee shops, and restaurants, to gain knowledge about the Korean culture and its traditions, such as gastronomy, art, music, and dance. What differentiates these visitors from others is their common interest in drama, animes, and K-Pop music, which have spread globally, supported by the South Korean government.

Interviews with participants indicated K-Pop enthusiast visitors communicate through social media networks, where they can follow other enthusiasts, get tips about Korean products and cultural spaces, and learn the Korean language. They stroll through the area searching for new experiences in the urban space, motivated by their social media networks formed by the followers of Korean culture. This global phenomenon, which extends beyond music, has created a sense of unity through video-clips, presentations, and choreographies (Santos, 2016). This mobility of images and sonorities impacts the youth in such a way that many of them look for these traditional cultures, expressed by arts, gastronomy, language, and religion, under reconfigured ways – in this case through Korean culture in Bom Retiro. Furthermore, managers of the cultural entities create events within Bom Retiro to promote Korean culture. A Cultural Centre Director described Korean culture as a product in itself. This is further evidenced by the fact that these events are sponsored by the Korean government and global Korean corporations (e.g. Hyundai and Life's Good Electronics).

The Korean wave typifies the images that 'became commodities and (are) operating according to their own autonomous logic within a chain of free-floating signifiers' (Gotham, 2002: 1737). In this way, Korean entrepreneurs have the opportunity to promote their culture as a business strategy and, at the same time, to valorise it. In this context, the Korean ethnic legacy takes part in the development of Bom Retiro tourism through the commodification of global process, thus creating consumption spaces at the local level, and consequently, attracting a diverse set of people interested in its ethnic aspects. In such a way, the existing social, cultural, physical, and aesthetic scenarios from Korean ethnicity have created conditions for more relevant tourism development. The spatial overlaps among tourists, visitors and locals brings about an interesting mix, promoting vitality and diversity in this contemporary city (Allis & Vargas, 2015).

Normally, the visitors that tour Bom Retiro are interested in its cultural spaces (museums, cultural centres, etc.), heritage (Zone A) and the popular street shops (Zone B). Its location and easy access allow visitors who are walking or using the subway (located in the Zone A) to access the most well-known district tourist attractions. Therefore, it can be considered that some spaces that are less associated with standard tourism (Zone C) require incidental visibility as a secondary motivator for tourists, who search for attractions that represent both global and local trends.

Although a part of the Korean community preserves its traditions in the ethnic enclave that was shaped in Bom Retiro, the Brazilian economic situation forces them to search for alternatives. If neolocalism refers to a movement for restoring the singularity and authenticity of communities (Flack, 1997), the Korean cohesiveness focused on maintaining their culture can be considered a form of neolocalism. It occurs both within the community that aims to grow business and gain public recognition, as well as by the visitors who are interested in exploring Korean culture.

Neolocalism: A new approach to urban tourism?

Neolocalism refers to the combination of local resources with a community's sense of place. In such way, the Korean legacy in Bom Retiro represents the adoption of the iconography and symbols that express ethnicity. The effort to present cultural particularities has attracted a broader audience interested in ethnic traditions and references. The Old World image, in conjunction with contemporary marketing techniques (Mathews & Patton, 2016), has resulted in elements of neolocalism that differentiate the destination and promote a connected consciousness of the locality (Flack, 1997). Tourists become interested in the Korean ethnic legacy and its authenticity since most entrepreneurs preserve, in a way, their traditions. In a globalised world, the ethnic places are often seen as emblematic transnational spaces, embedding the cultural and economic globalisation process of the cities (Lin, 1998) and giving the singularities of Korean culture the potential to attract external audiences.

Looking into the Bom Retiro tourism dynamic and understanding neolocalism as the desire of visitors to reaffirm the place's singularity and authenticity (Schnell & Reese, 2014), the Korean ethnic legacy is a counterpoint to the mass tourism attractions in Bom Retiro (shopping and heritages). In this way, the culture of these migrants (and their descendants) sustains a sort of resistance to economic and urban landscape homogenisation (Holtkamp et al., 2016). Moreover, the growth of the neolocal movement provides an opportunity to hold at bay the globalisation process, which tends to dilute particularities of place and threaten authenticity (Wright & Eaton, 2018).

It is noticeable that tourists and visitors interested in museums, historic buildings, and shopping experiences contribute to the solidification of tourism within the Korean ethnic legacy. This is shown by the growing interest in Korean culture from tourists and other visitors. Therefore, it deals with the advent of the aestheticization of daily life, despite the convention qualifiers of the tourism definition.

Conclusions and future developments

Globalisation has significantly shaped tourism over the years. Deindustrialisation has driven urban economies into the services sector in major global cities. In these circumstances, cities provide attractive spots for locals, even when these spaces are familiar (Allis, 2012, 2014; Spirou, 2011; Zukin; 1996), such as the Train Station Luz in São Paulo that receives visitors, tourists, and workers. In many situations, locals consume contemporary urban experiences in a similar way as tourists, enjoying the same activities, playing similar functions, and performing similar experiences within the same spaces (Edwards et al., 2008; Maitland, 2010; Selby, 2004).

John Urry, Mimi Sheller (2006) and other researchers who are anchored in the mobilities field (Gale 2008, 2009; Hannam, 2009) have called attention to what was termed the 'end of tourism', when tourism specificities are blurred with other forms of production, consumption, and mobility (Gale, 2009: 122). This arrangement is driven by the fragmentation of consumption choices, the introduction of new information technologies, and changes in the labour division (Hall & Rath, 2007). Bom Retiro could be described as a different place within the city, where certain 'spatial behaviours' (Hayllar et al., 2010: 139) are represented by cultural heritage and ethnic economies. This set of elements is complex, shaping the experience in a way that encourages all actors (Korean, visitors, tourists, locals, workers) to interact with attractions, touristic and non-touristic infrastructures as a way to bring cultures together.

In the end, the scenario of the touristic dynamic of Bom Retiro, which involves the Korean ethnic legacy and existing tourism, despite their particularities, shows that some factors are preponderant for the development of local Korean culture that triggers various types of mobilities, for both locals and non-locals. As the conceptual boundaries blur, so do the geographical ones. Engaging in certain representations and embodiments, visitors to Bom Retiro perform certain spatial behaviours that would hardly be framed as purely leisure or conventional tourism.

Regarding a future agenda for policy-makers and communities, the identification and discussion of these issues are of strategic importance, as they highlight opportunities, as well as potential threats, to overlapping dimensions of daily urban life. Assuming this is not a topic only for the tourism agenda, one has to realise that new urban dynamics rooted in ethnic values require attention from several fields (culture, education, economics, urban planning, and so on). Ultimately, if we head towards the 'end of tourism', de-differentiation in the daily lives of huge metropolis dwellers is gradually realised, combined or hybrid solutions, in terms of public policies, should apply.

References

Allis, T. (2012) Projetos urbanos e turismo em grandes cidades: o caso de São Paulo, PhD thesis, University of São Paulo.

Allis, T. (2014) Viajantes, visitantes, turistas... em busca de conceitos em um mundo urbano. *Caderno Virtual de Turismo,* **14**(1), 23-38.

Allis, T. & Vargas, H.C. (2015) Turismo Urbano em São Paulo: reflexões teóricas e apontamentos empíricos. *Revista Turismo em Análise,* **26**(3), 496-517.

Antônio, B. & Araújo, J.R.C. (2019) A diáspora coreana: o caso brasileiro. *Confins. Revue franco-brésilienne de géographie. Revista franco-brasileira de Geografia,* **39**.

Baldissera, L.M., Bahl, M. & Nitsche, L.B. (2017) Legado étnico holandês como atrativo turístico na comunidade de Castrolandam, Paraná, Brasil, *Revista Confluências Culturais,* **6**(2), 59-72.

Bell, D. & Valentine, G. (1997) *Consuming Geographies: We are where we eat*, 1st edn, New York, NY: Routledge.

Bitelli, F.M. & Bastos, S.R. (2018) Hospitalidade na cidade: as manifestações culturais como uso do espaço público. *Turismo Visão e Ação*, **20**(3), 460-473.

Bocci, D.S. (2009) Bairro da Liberdade e a imigração japonesa: a ideia de bairro Japonês, *Cordis: Revista Eletrônica de História Social da Cidade*, (2).

Brasil. (1988) Constituição da República Federativa do Brasil. Artigo 216, Seção II da Cultura, available at https://www.senado.leg.br/atividade/const/con1988/CON1988_05.10.1988/art_216_.asp. (Accessed: 2/07/2019).

Carvalho, M.A. (2011) Cidade global, destino mundial: turismo urbano em São Paulo, PhD thesis, University of São Paulo.

Chamber of Shopkeepers. (2017) São Paulo. Available at http://www.bomretiro.cdl-sc.org.br/institucional/a-cdl/. (Accessed: 13/04/2019).

Chi, J.Y. (2016) O Bom Retiro dos coreanos: descrição de um enclave étnico, PhD thesis, University of São Paulo.

Choi, K.J. (1996) Imigração coreana na cidade de São Paulo, *Revista do Instituto de Estudos Brasileiros*, **40**, 233-238.

Collins, J. (2007) Ethnic precincts as contradictory tourist spaces, in C.M. Hall & J. Rath (eds.), *Tourism, Ethnic Diversity and the City*, London: Routledge, pp. 85-104.

Cordeiro, H.K. (2002) A 'cidade mundial' de São Paulo e o complexo corporativo do seu centro metropolitano, in M. Santos, M.A.A. de Souza, F.C. Scarlato, & M. Arroyo (eds.), *O novo mapa do mundo – Fim de século e globalização*, Hucitec & Anpur, 4th ed, pp. 318-331.

Cymbalista, R. (2007) A comunidade boliviana em São Paulo: definindo padrões de territorialidade. *Cadernos Metrópole*, **17**, 119-133.

Companhia Paulista de Trens Metropolitanos (CPTM) (2018) Relatório Integrado da Administração 2018. Companhia Paulista de Trens Metropolitanos, São Paulo.

da Cruz Paiva, O. (2011) Territórios da migração na cidade de São Paulo: Entre a afirmação e negação da condição migrante, *Ideias*, **2**(1), pp. 13-30.

Dertônio, H. (1971) *O bairro do Bom Retiro* (Vol. 9). Prefeitura Municipal, Secretaria de Educação e Cultura, Departamento de Cultura.

Edwards, D., Griffin, T. & Hayllar, B. (2008) Urban Tourism Research Developing an Agenda, *Annals of Tourism Research*, **35**(4), 1032-1052.

Fainstein, S.S. & Gladstone, D. (1999) Evaluating urban tourism, in D. Judd & S. Fainstein (eds.), *The Tourist City*, New Haven, CT: Yale University Press, pp. 21-34.

Flack, W. (1997) American microbreweries and neolocalism: "Ale-ing" for a sense of place, *Journal of Cultural Geography*, **16**(2), 37–53.

Florida, R. (2011) *A ascensão da classe criativa: e seu papel na transformação do trabalho, lazer, comunidade e cotidiano*, Porte Alegre, Rio Grande do Sul, Brazil: L&PM Editores.

Fonseca, M.C.L. (2009) Para além da pedra e cal: por uma concepção ampla do patrimônio cultural, *Memória e patrimônio: ensaios contemporâneos, 28.*

Freitas, P.T. & Baeninger R. (2010) Cidade e Imigração: Origens e Territórios da Imigração Boliviana e Coreana para a Cidade de São Paulo, *População e Cidades: subsídios para o Planejamento e para as Políticas Sociais, Campinas,* **1**, 137-169.

Gale, T. (2008) The end of tourism, or endings in tourism, in P. Burns & M. Novelli (eds.), *Tourism and Mobilities: Local-Global Connections*, Wallingford, Oxfordshire: CABI, pp. 1-14.

Gale, T. (2009) Urban beaches, virtual worlds and 'the end of tourism'. *Mobilities,* **1**(4), 119–138.

Genestreti. (2015) Exposição de Ron Mueck bate recorde da Pinacoteca de SP, Portal G1. Available at https://www1.folha.uol.com.br/ilustrada/2015/02/1593438-exposicao-de-ron-mueck-bate-recorde-da-pinacoteca-de-sp.shtml (Accessed: 25/07/2019).

Global Destination Cities Index. (2018) *Welcome to the Engagement Bureau.* Available at https://newsroom.mastercard.com/latin-america/pt-br/digital-press-kits/global-destination-cities-index-levantamento-anual-realizado-pela-mastercard/. (Accessed: 13/07/2019).

Gotham, K.F. (2002) Marketing Mardi Gras: Commodification, spectacle and the political economy of tourism in New Orleans, *Urban Studies*, **39**(10), 1735-1756.

Gotham, K.F. (2005) Tourism gentrification: The case of New Orleans' Vieux Carre (French Quarter), *Urban Studies*, **42**(7), 1099-1121.

Hall, C.M. & Rath, J. (2007) Tourism, migration and place advantage in the global cultural economy, in C.M. Hall & J. Rath (eds.), *Tourism, ethnic diversity and the city*, London: Routledge, pp. 19-42.

Hannam, K. (2009) The end of tourism? Nomadology and the mobilities paradigm, *Philosophical Issues in Tourism*, **37**, 101-116.

Harvey, D. (1989) *The Urban Experience*, Baltimore: Johns Hopkins University Press.

Hayllar, B., Griffin, T. & Edwards, D. (2010) *City Spaces-Tourist Places*, London: Routledge.

Holtkamp, C., Shelton, T., Daly, G., Hiner, C.C. & Hagelman III, R.R. (2016) Assessing neolocalism in microbreweries, *Papers in Applied Geography*, **2**(1), 66-78.

Kara-José, J.B. (2007) *Políticas culturais e negócios urbanos: a instrumentalização da cultura na revitalização do Centro de São Paulo: 1975-2000*, Annablume, Fapesp, São Paulo.

Lin, J. (1998) *Reconstructing Chinatown: Ethnic enclaves and global change*, Minneapolis, MN: University of Minnesota Press.

Lincoln, Y.S. & Guba, E.G. (1985) *Naturalistic Inquiry*, Thousand Oaks, CA: SAGE.

Lune, H. & Berg, B.L. (2017) *Qualitative Research Methods for the Social Sciences*, 9th edn, Essex: Pearson Education Limited.

Maitland, R. (2010) Everyday life as a creative experience in cities, *International Journal of Culture, Tourism and Hospitality Research*, **4**(3), pp 176–185.

Mathews, A.J. & Patton, M.T. (2016) Exploring place marketing by American microbreweries: neolocal expressions of ethnicity and race, *Journal of Cultural Geography*, **33**(3), 275-309.

Menezes, G. (2019) Onde comer bem no Bom Retiro, *Veja* São Paulo online. Available at https://vejasp.abril.com.br/comida-bebida/onde-comer-bem-no-bom-retiro/(Accessed: 24/12/2019).

Pinacoteca. (2018) Associação Pinacoteca de Arte e Cultura. Planilha de Público, São Paulo.

PLATUM. (2015-2018) *Plano Nacional de Turismo: Cidade de São Paulo*. São Paulo Turismo, São Paulo.

Plummer, R., Telfer, D., Hashimoto, A. & Summers, R. (2005) Beer tourism in Canada along the Waterloo–Wellington Ale Trail, *Tourism Management*, **26**(3), 447–458.

Póvoa, C.A. (2007) A territorialização dos judeus na cidade de São Paulo-SP: a migração do Bom Retiro ao Morumbi, PhD thesis, University of São Paulo.

Rath, J. (ed.) (2007) *Tourism, Ethnic Diversity and the City*, 1st edn, New York, NY: Routledge.

Santos, T.H. (2016) Idols em imagens e sons, fãs em re-ação: uma etnografia da prática musical do K-pop em *São Paulo*, PhD thesis, University of São Paulo.

São Paulo Cosmópolis. (2017) *Imigrantes em São Paulo : diagnóstico do atendimento à população imigrante no município e perfil dos imigrantes usuários de serviços públicos* / organizado por São Paulo Cosmópolis ; coordenado por Deisy Ventura, Feliciano Sá Guimarães e Rossana Reis. São Paulo : IRI-USP, 170 p.

Sassen, S. (1991) *The Global City: New York, London, Tokyo*, Princeton, NJ: Princeton University Press.

Selby, M. (2004) Consuming the city: conceptualizing and researching urban tourist knowledge, *Tourism Geographies*, **6**(2), pp. 186–207.

Schnell, S.M. & Reese, J.F. (2014) Microbreweries, place, and identity in the United States, in M. Patterson & N. Hoalst-Pullen (eds.), *The Geography of Beer: Regions, Environment, and Societies*, Dordrecht: Springer, pp. 167-187.

Sheller, M. & Urry, J. (2006) The new mobilities paradigm, *Environment and planning A*, **38**(2), 207-226.

Spirou, C. (2011) *Urban Tourism and Urban Change: Cities in a global economy*, London: Routledge.

Timothy, D. (2002) Tourism and the growth of urban ethnic islands, in M. Hall & A. Williams (eds.), *Tourism and Migration*, Dordrecht: Springer, pp. 135-151.

Truzzi, O. (2001) Etnias em convívio: o bairro do Bom Retiro em São Paulo, *Estudos Históricos*, **28**, 143-166.

Truzzi, O. & Sacomano Neto, M. (2007) Economia e empreendedorismo étnico: balanço histórico da experiência paulista, *Revista de Administração de Empresas*, **47**(2), 1-12

Vargas, H.C. (2000) Turismo nos grandes centros urbanos, *IV Encontro Nacional de Turismo, Joinville, SC.*

Venceslau, P. (2017) Doria quer revitalizar e mudar o nome do bairro para 'Bom Retiro Little Seul', *Estadão online.* Available at https://sao-paulo.estadao.com.br/noticias/geral,doria-quer-revitalizar-bom-retiro-e-mudar-nome-para-bom-retiro-little-seul,70001736986. (Accessed: 14/12/2019).

Williams, A.M. & Hall, C.M. (2000) Tourism, migration, circulation and mobility, in A.M. Williams & C.M. Hall (eds.), *Tourism and Migration,* Dordrecht: Springer, pp. 1-52.

Wright, W. & Eaton, W.M. (2018) Representing rurality: Cider mills and agritourism, in S.L. Slocum, C. Kline, & C.T. Cavaliere (eds.), *Craft Beverages and Tourism, Volume 2,* Champagne, IL: Palgrave Macmillan, pp. 65-82.

Yang, E.M. (2011) A geração 1.5 dos imigrantes coreanos em *São Paulo:* identidade, alteridade e educação, PhD thesis, University of São Paulo.

Zukin, S. (1996) *The Cultures of Cities,* Hoboken, NJ: Wiley-Blackwell.

8 Neolocalism and Transportation: Biocultural Conservation of the Saen Saeb Khlong in Bangkok

Christina T. Cavaliere, Vorakarn (Kluay) Chalermchaikit, Yixuan (Cindy) Cheng, and Akruthai (Amy) La

Abstract

The post-modern world is comprised of a set of confluences that include a mixture of globalised, and therefore increasingly homogenised, experiences. Nevertheless, the tourism sector relies on the uniqueness of place as a core driver of the phenomenon. This chapter focuses on the specific context of the Saen Saeb canal (khlong) in Bangkok, Thailand. The historical significance of the khlong system, the current context of khlong riverboat transportation and the possibility of future tourism development is discussed. In addition, this research brings forth deeper questions of equity and distribution of positive and negative impacts of neolocal transportation development for residents and tourists alike. It includes suggestions regarding the importance of stakeholder involvement to create plans for khlong conservation and development to serve residents and increasing numbers of tourists. The chapter presents an application of the lens of neolocalism as explored through issues of heritage transportation, social capital, biocultural conservation, and sense of place. Issues of biocultural conservation through collaborative partnerships and good destination management strategies for long-term sustainable tourism planning are also proposed.

Introduction: neolocal transportation, social capital and sense of place

This chapter aims to explore the interrelationship between heritage transportation, social capital, biocultural conservation, and sense of place through the application of the lens of neolocalism within the specific context of a river canal (known as *khlong*) in Bangkok, Thailand. The post-modern world is made up of a set of confluences that include an ever-increasing rapid mixture of globalised, and therefore increasingly homogenised, experiences. Yet, for the tourism sector, the uniqueness of place is a core driver of the phenomenon. We travel for many reasons, and often it is to experience differences and 'the other'. Concurrently, for locals, tourism can serve as a driver of capitalistic homogenised development and can result in the loss of heritage and sense of place.

Neolocalism, as a philosophical building block, seeks to address conceptualisations of a renewed sense of place. Schnell (2013a: 624) describes this conceptualisation as a 'countermovement...the conscious fostering of local connections, identities, and economies'. Shortridge (1996) is credited as first identifying neolocalism as a trend and as an intentional search for identity in a place in an 'increasingly rootless society' (Flack, 1997: 38). Since then, the idea has transmuted and changed, and it is now noted that people are forming national and international networks of neolocals, 'mutual support organizations for their mutual interests in preserving distinctive places' (Schnell, 2013a: 71).

The act of travel and the interest in visiting new places are ever increasing globally. The United Nations World Tourism Organization (UNWTO) predicts that from 2010 to 2020, there will be a growth in world tourism of 3.8% annually (UNWTO, 2011). This growth of the tourism sector is strongly evident in Thailand. The Tourism Ministry of Thailand forecasts that inbound visitors will reach approximately 41.1 million this year (Bangkok Post, 2019). This growing trend is largely attributed to the expansion of a global middle class, the improvement of better infrastructure with improved accessibility to destinations, and a wider variety of traveling options that are more suitable to a range of travellers' budgets (OECD, 2018).

Changes in conscious consumption also involve notions of personal awareness of the relationships between individuals, the environment, and the place, and thus, neolocalism engages with more personal experiences of the nature of local places. Schnell (2013b: 625) argues that this type of new awareness to local connections are understood as more 'authentic, and are an increasingly important part of place-creating narratives'. With this increase in demand, there is also a change in travellers' preferences and includes those who seek more authentic and localised experiences. Today's tourists are becoming more selective and want to travel more 'off the beaten track' to places that are less explored and advertised by the media. Therefore, visitors want to experience the authenticity of the local culture ranging from local gastronomy to participating in local traditional celebrations and activities (Revfine, 2019). These travellers and tourists are seeking to experience a sense of place. According to Wang and Xu (2015: 241), 'sense of place can be described as a compilation of meanings, knowledge, attachment, commitment and satisfaction that an individual or group associates with a particular place'. In addition, local tourism trends, or 'staycations', are becoming popular among people who do not want to travel far from home but still seek to experience new places and activities. Local tourism can involve travellers seeking involvement with local community, and therefore can raise awareness and enable visitors to appreciate and better understand local authenticity in terms of place, people, culture, food, and history (Majumdar, 2018).

This chapter explores notions of ways in which neolocalism can serve as a form of biocultural (biological and cultural) conservation. Through the reinvigoration of interest in local tourism, the use and protection of historical transportation routes can potentially be both conserved and utilised by locals and tourists alike. Yet, according to Ujang and Zakariya (2015: 710), evidence indicates that 'urban regeneration within traditional settings has diminished the place meanings of transformed and or newly constructed places'. This is relevant to the context of the khlongs in Bangkok as many of the canals have been filled in with earth to create more roadway space over time due to modernisation efforts of transportation infrastructure in the megacity. For example, Parnwell's (2007) study utilises the conceptualisation of neolocalist revivalism as related to rural communities in northeast Thailand. He contends that this approach is a strategy to combat the ever-present effects of relentless modernisation and development on society, the economy, and the environment by reviving a variety of social institutions and associational practices that underpin social capital. His empirical case study in northern Thailand demonstrates that 'social capital can (re)emerge from the ashes of modern development to become the lubricant of (neo)localist efforts to build alternative, community-centered and ostensibly sustainable forms of development' (p. 990). However, social capital must be understood, and he claims that often it remains invisible. According to Parnwell (2007: 991), social capital is positioned within 'social structures, it centres on the networks, connections, collaborations, interpersonal relationships...of a given society or community'. This chapter contends that the khlongs serve as social capital and, if made evident as such, they can then be more clearly understood as valuable and, therefore, warrant further study and conservation. Thus, this chapter will proceed with a presentation of a brief history of the khlongs, the methodological approach of the study of one specific khlong, and the implications of further linking khlong transportation routes to the study of neolocalism.

Background of khlongs: biocultural conservation of historic waterways

Bangkok has been recognised as the 'Venice of the East' because during the period of the 19th century many river canals were dug as offshoots to the main Chao Phraya River that runs through the city. Today, the historic waterways that still intersect through the capital serve as the living cultural and historical remainders of this heritage (Bodry, 2012). A *khlong* is a channel of water, and historically they served as the lifeblood of the general public (Jiarakul, 2016). Khlongs were also historically built as approaches to security in times of threats of outside intruders. Currently, there are 1,161 khlongs in Bangkok (Bangkok Post, 2018). However, today, land transportation has outpaced canal usage, although the canals have functioned as the main transportation mode in previous generations (Kritiyutanont et al., 2016). As aforementioned, in modern times, many of these historical canals have been filled in due to urban development.

This chapter explores conceptual notions of neolocal transportation within one khlong known as the Saen Saeb. Its community is varied and diverse and includes prominent groups of Thai-Muslims in Bangkok. The khlong includes various communities of people who live along the banks of the Chao Phraya River, including the areas of Mahanak khlong and Saen Saeb khlong. These local communities' livelihoods have been closely linked to these khlongs in terms of trading, transport, and agriculture (Phaithayawat, 2015). However, over time, several factors involving modernisation, urban development, population growth, and changes in livelihood sources have drastically changed people's relationship to the waterways. In the present day, most khlong are used for drainage and sewage disposal. However, Saen Saeb and Phasi Charoen khlong are still being used for the purpose of transportation (Jiarakul, 2016). These last remaining modern routes are critical to the communities that reside in this subregion for various reasons, including the fact that they serve as very fast and cheap transportation for the needs of daily life. Bangkok's roadways are very congested and short distances can take a very long time to transverse in both private and public transportation options. In addition, many of these modern transportation modes are expensive and not feasible for lower income community members.

In addition to the important utilitarian use by local communities, these khlongs can be further developed for tourism purposes. There are three different kinds of boat operations that provide varying services in Bangkok. These include the express boats for traveling along the Chao Phraya River, the ferry boats for crossing the river, and the long-tail boats for commuting into the khlongs (Iamtrakul et al., 2017). Prakraiwan's (2016) study concluded that there is the potential for the development of a cultural tourism route within this area of Bangkok. The khlong area has significant cultural resources, such as temples, various ancient communities, and unique wooden houses (that are also quickly disappearing due to urban development). Furthermore, the community has changed significantly due to an increase in population and, therefore, traditional residential homes are disappearing to make way for new development (Suwandee et al., 2013).

There are several opportunities for tourism development, as related to the khlongs, that include access to services, such as accommodation, local food, and transport convenience (Giriwati et al., 2013). In addition, floating markets are popular among tourists as they represent historic cultural livelihoods. For example, Batra's (2014) study explored the perspectives and relationship between visitors and locals in a case study of the Taling Chan floating market in Bangkok. His research determined there is a significant linkage between this canal community and the economic impacts of tourism in terms of poverty reduction and distribution of earnings to locals.

Specifically, khlong transportation has been identified as having potential to be promoted as a tourist activity, however, it would require additional incorporation into the municipality and national tourism strategies (Inland Water Transport Development Company, 2018). Of important consideration as related to khlongs and tourism product development, is the impact this can have on the local community members who are currently utilising this very affordable and convenient mode of transport for daily life (Iamtrakul et al., 2018).

Khlongs and Bangkokian livelihoods

Transportation within Bangkok is an incredibly challenging and critical aspect of daily life for both locals and tourists alike in this urban context and includes faster options via water transport on the Chao Phraya River. According to Bangkok's Department of Drainage and Sewerage (2019), there are up to 1,161 khlongs and the two primary khlongs are Saen Saeb and Phasi Charoen. Saen Saeb khlong was constructed by King Rama III in 1837 for military services' transportation (Khaosod, 2016). However, in modern times, Saen Saeb khlong has been transformed into the main waterway for many local people. Saen Saeb Express Boats, the oldest express boat transportation company, has been in operation since October 1990. The boats operate every day and offer two lines and 29 stops along 18 kilometres from downtown to the east side of Bangkok with a cost of approximately 10 to 20 baht, depending on the distance (Transit Bangkok, n.d.). Due to the speed and convenience of khlongs and the demand of passengers in the West of Bangkok, Phasi Charoen Express Boats, launched in 2016, offers 15 stops along 11.5 kilometres connecting to the sky train (BTS-Bangwa station).

Currently, ground transport (including motorcycles, taxis, tuk-tuks, vans, buses, and private cars) congestion has escalated in the densely populated city of approximately 8.2 million inhabitants (World Population Review, 2019). Long and disruptive traffic jams are part of the way of life in the city. Therefore, to minimise these major traffic problems, rapid trains have been developed, such as Bangkok Transit System (BTS/sky train) in service since 1999 and Metropolitan Rapid Transit (MRT) in service since 2004, with multiple transit stations in areas such as Sathorn, Sukhumvit and Siam. Locals and tourists prefer utilising BTS and MRT because it is convenient with planned arrival and departure times, however the cost is high in comparison to other public transportation options. According to BTS (2019), the tariffs (based on distance) range from 15-59 baht and for MRT from 21-42 baht (BEM, 2018). In comparison, public buses cost from 8-25 baht (BMTA, 2019). Although the tariffs are less, the challenges involve the extraordinary amount of time it may take during rush hour and unpredictable bus schedules. Thus, many people try to avoid taking road transport as much as possible. The typical minimum wage of a Thai employee is approximately 300 baht per day. A daily roundtrip BTS/MRT ride will cost approximately 100 baht, a significant portion of their salary. Hence, pricing is a barrier in transport choices for locals. Khlong transport is an ideal option due to the high speed, no traffic, and is accessibly priced. According to the Thai Marine Department (2019), Saen Saeb express boat costs from approximately 9-19 baht depending on the distance. Hence, khlong transportation has benefits for local passengers for convenience and for tourists regarding exposure to authenticity and sense of place.

Canal tourism development in other regions

Previous international studies note that canals have experienced many transformations over the years as alternative land transport modalities have improved. Recent studies indicate a trend that there is an increase in the use of canals for leisure and tourism activities (Kaaristo & Rhoden, 2017). For example, in the UK, there are over 2000 miles of inland waterways that are managed by different navigation institutions. It is reported that the usage of leisure boats on canals has increased to 19.5 million visitors in 2014, which indicates the great potential for canal tourism development (Kaaristo & Rhoden, 2017). Kaaristo and Rhoden's (2017) research demonstrates that 'spatial practices, temporal structures, and search for ontological security are essential elements of …co-creating the emergent boatspace, where physical and social space are negotiated between various actors' (p. 78).

Tang and Jang (2010) examined the transformation of the New York Canal System from a commercial transportation waterway to a tourist destination providing sightseeing and heritage interpretation. Management authorities have implemented a set of diverse marketing strategies to promote the development (Tang & Jang, 2010). This case study also indicates success via integrated funding strategies involving the state government, local, regional, non-profit, and private organisations and enterprises, as well as local communities (Tang & Jang, 2010). The study also identifies various challenges, including the lack of a unified image as a tourism destination and the need for local communities along the canal to reduce conflicts and increase cooperation. In addition, negative impacts from water pollution, a decrease in flora and fauna species, and solid waste disrupt the balance of the ecosystem, and cause issues for local populations that require monitoring and evaluation for the sustainable development of canal tourism (Tang & Jang, 2010).

The Wang (2012) study explores the appropriate solutions and strategies to manage The Grand Canal in China that passes through 35 cities. In ancient China, the Grand Canal was utilised for military and economic functions. The canal flows through several provinces and integrates the South-North Water projects and, therefore, the management of The Grand Canal requires cooperation among a wide range of institutions, including the Heritage Department, Transport Department, Water Resource Department, and Environment Protection Department (Wang, 2012). Wang's research identifies that there is conflict between various authorities due to different missions and interests and, yet, better cooperation is essential for good management. In order to successfully manage and conserve The Grand Canal, a collaborative approach is suggested to resolve conflicts and achieve a shared vision via a joint decision-making process that involves various stakeholders. Wang (2012) highlights that the Grand Canal remains primarily a transport system, and thus 'heritage must remain ancillary to the need expressed by most planners to improve it as a transport route' (p. 39). However, the Grand Canal has been recognised as a vast cultural site, and the process of being included in the UNESCO World Heritage List requires that special laws and regulations must be implemented as soon as possible.

Shipley's (1999) study aims to explore the Welland Canals in the Niagara region as a tourism development destination. Niagara Falls is a famous natural attraction and accommodates large numbers of tourists. Welland Canals also includes cultural tourism and preservation of this heritage is supported by a strong local community movement to maintain the destination, as understood by the researcher after six years serving as both an observer and participant who worked as a citizen volunteer and planner for community development. A community group, the Welland Canals Preservation Association (WCPA), was created to plan and maintain the destination, and to enhance community engagement. Shipley identified five successful elements of canal management: an appreciation of the resource; citizen participation; government leadership; inter-community cooperation; and the control of vested interests. First, the appreciation of this resource as a cultural/heritage resource impacts canal structure and should be managed by different levels of government and organisations. Second, citizen participation encourages a far-sighted vision and includes citizens, along with the WCPA, as it is important to represent perspectives of community members in the process of tourism planning and development. Third, government leadership enhances effectiveness of strategic planning and legislation affecting infrastructure and initiative construction. Also, it is crucial for the government to recognise the importance of potential tourism attractions and take appropriate measures to manage and preserve them. Fourth, inter-community cooperation allows canal development among various stakeholders. Finally, the control of vested interests fulfils a collaborative model in community tourism because the local originators of canals know their plan and community involvement.

Neolocalism emphasises place-specific efforts by local communities to develop and enhance self-reliance, sustainability, and social harmony based on the interplay between social resources and natural resources (Parnwell, 2007). As Curtis (2003) states, the process of neolocalism focuses on the development of a moral and social economy to protect and maintain the local resources and environment. Under the context of tourism, neolocalism may indicate a potential approach where sustainably developing and exploring tourist attractions based on local natural and cultural capital enhance a sense of place and local identity. This previous literature indicates that canal tourism can change a destination's image because it highlights cultural heritages and enhances tourists' access to cultural and recreational activities. Therefore, understanding tourism as related to the local daily activities of a destination is crucial. In addition, based on the concept of neolocalism, the combination of local daily routines and tourism development can be further explored in terms of community involvement.

Methodology

This study focuses on neolocalism, social capital, and khlong transportation using a qualitative approach. According to Barnham (2015), a qualitative method is based on in-depth understandings through observations and engagement. Therefore, participant observation was implemented with an aim to participate in the lives and experiences of the subjects that enabled the researchers to collect data by observing and experiencing what was happening (Gill & Johnson, 2002). The purpose of participant observation is to explore the meaning behind a specific phenomenon (Saunders, 2011). According to Gill and Johnson

(2002), there are four roles that researchers can play in participant observation research. The roles are complete participant, complete observer, observer as participant, and participant as observer. The complete participant means researchers attempt to become a member of communities or groups in which they conduct the research and do not reveal the true propose to the subjects in order to collect deep data.

This research project was an integral part of a tourism planning doctoral course being taught by the first author in Bangkok. It is important to note that this study was encompassed within the context of a doctoral course and thus the researcher participants had many hours of interaction and engagement together as a group focused on aspects of tourism planning. This research was designed in three parts comprised of a pre-field planning session, the data collection process in the field, and then multiple post-trip analysis discussions. The group determined during the pre-trip planning sessions that the data collection was to involve a semi-structured two-fold approach that included the use of our five senses as engagement (sensual engagement) and by focusing on ten observation points. Sensual engagement allowed the researchers to fully recognise 'that we are embodied beings that encounter and interact through the senses with our everyday world' (Duffy & Mair, 2018: 50). During the preparation for the field work, these observational frameworks were utilised in order to focus the researchers on specific topics to ensure some commonality of attention when collecting the data. The pre-field planning also included preparation activities, where the group explored their own assumptions and expectations of the field experience on Saen Saeb khlong. The actual field work and data collection involved a full day on the khlong. The researchers also collected personal observation notes outside of the 10 observation points and included photography as a method of observation. Consequently, researchers had individual memories to share with the research group regarding personal thinking and experiences. Analysis occurred after the field visit, when the research group met to compare and discuss the data collected.

In July of 2018, the four researchers served as observational participants spending a full day riding the express boat roundtrip from the university to downtown on the Saen Saeb khlong along with community members and tourists (see Image 8.1). The data was collected from what we experienced from two boat rides (a complete round-trip ride of 80 minutes on the khlong) and from observations conducted while in the khlong station terminals. During the time spent in the khlong stations waiting for the boats on the way to downtown and back, we discussed our experiences and utilised sense information from what we saw, smelled, heard, tasted, and touched. By utilising our five senses, visual observation of other passengers, observable cultural heritage along the shore, and community and infrastructure design, we collected copious notes. After the day collecting field observations, the group met to discuss the experiences, compare field notes, and begin to structure data analysis. We compared notes and the data via discussion, and the data was also constituted by comparing photographs of the specific images of each context and was based on story telling (Jansen, 2015).

Image 8.1: Saen Saeb khlong express boat. Photo by Christina T. Cavaliere, 2019.

Therefore, we set out with the predeveloped semi-structured observation points, including our sensual experiences and the ten observation points, to collect data during an eight-hour roundtrip voyage to downtown and back from the university. During the field visit, researchers recorded what they observed and experienced with an understanding that data collection and analysis would be carried out in participant observation research. The four researchers originated from different countries including Thailand, China, and the USA and, therefore, had different pre-field observation expectations. Upon returning from the field observation and data collection, the researchers compared and discussed our sensual experiences and the ten observational points. The comparison analysis of the data points resulted in the development of a Strengths, Weaknesses, Opportunities, and Threats (SWOT) analysis (see Table 8.1).

Results

First, we noted the condition of the boats and found that the boats were older and not in good condition. For instance, there were big holes in the bottom of the boat, which passengers could fall into if they were not careful. The experience was different from what we expected because we assumed that the boat condition would be better.

Second, we noted the condition of the khlong stations. Because the express boat is the cheapest transportation, the stations were not in good condition. The infrastructure was very old and required maintenance. Specifically, we noted that there were no predetermined spaces or areas for passengers to queue, so everyone had to literally jump onto the boats quickly, which seemed very unsafe.

Third, the researchers noted the friendliness of the boat staff. We expected a moderate level of engagement because it depended on individuals' personalities. However, we found that many of the staff were polite and kind, except one incident when one staff member yelled at one of the researchers to hurry up during disembarkation of the boat.

Fourth, we observed the presence and usefulness of the khlong station signage and, as we had anticipated, most of the signage was provided in Thai language. Although there was some limited English that appeared on signage closer to downtown, there was far less information available in English than in Thai. Moreover, the sizes of the maps were generally very small and, therefore, made it difficult for passengers to see.

Fifth, we observed aspects of safety and security during the experience. The researchers had planned to be very careful while getting on and off the boats because we had heard that some passengers had been injured when they rode the express boats. There were personal flotation devices or lifejackets tucked into the top of the boat, but it did not appear that there were enough for each passenger on the boat. In addition, because the water is highly polluted, passengers are provided plastic sheets to protect them from splashing water. If a passenger were to fall into the water, there would be an emergency as they would need immediate medical attention due to toxic exposure. There were no instructions for any type of emergency situations. Furthermore, boats were not fully stopped for passengers to get in and out. The express boats slow down as they approached the dock and then the passengers were required to jump on or off board (see Image 8.2). It seemed unsecure because passengers could easily slip and fall without handrails etc.

Image 8.2: Challenges with disembarkation. Photo by Christina T. Cavaliere, 2019.

The sixth observation point included the views from the boat. The researchers experienced very beautiful natural and heritage scenery along the shore. We observed authentic wooden Thai houses (that are quickly disappearing due to modern urban development), commercial buildings, local people, religious temples, and sites of cultural infrastructure. For example, we observed the integration of Muslim and Buddhist communities, temples, local people, and the juxtaposition of both modern and historic buildings (see Image 8.3).

Image 8.3: View of Thai Buddhist temple (Wat) from Saen Saeb khlong. Photo by Christina T. Cavaliere, 2019.

The seventh point involved the observation of the authenticity of Thai livelihoods. The researchers observed authentic and historic Thai architecture, lifestyles (namely through food vendors and the exteriors of homes) and the reality of low socio-economic living conditions. Therefore, we felt we experienced a limited view of the reality of daily life of local people. There were only three or four passengers that appeared to be tourists (although this was not validated as we were not able to speak with them) however, it was clear that most of the passengers were local people.

The eighth observation point included aspects of accessibility of the boats themselves. As previously discussed, the researchers found that it was extremely difficult, if not entirely impossible, for passengers with mobility challenges, elderly people, and children to get in and out of the vessels.

The ninth aspect involved access to public restrooms. The researchers observed that there were only toilets in one big station that was closest to downtown. However, they were in poor condition because they were dirty and were emitting very strong smells and were not hygienic.

Lastly, we collectively decided to eat in the middle of the day as we were interested in experiencing local Thai food. One researcher suggested, from previous knowledge, that we visit a restaurant with Isan (North-Eastern Thai) cuisine. We did need to utilise a local tuk tuk taxi to access the restaurant from the khlong pier. This restaurant was interesting because it highlighted traditional Thai food and locally produced art, which added to the cultural experiences.

In conclusion, the research group collected predeparture expectations, a full set of observational field notes that were loosely structured around the use of our five senses, ten observation points, individual photographs, and personal narratives. These data were compared and analysed after the trip during research meetings and included explorations of our experiences from the structured notes and additional personal narratives of each researcher. The research group then collectively developed a SWOT analysis of the data based on data comparison and analysis.

Table 8.1: SWOT analysis of Saen Saeb khlong transportation development

Strengths

- Reasonable transportation prices
- Historic and modern public transportation
- Quick and efficient transportation time and routes
- Cultural exposure
- Local authenticity
- Religious exposure
- Access to local gastronomy
- Sightseeing for tourists
- Local jobs
- Unique transportation service

Weaknesses

- Funding challenges
- Carrying capacity of current services
- Lack of effective policies
- Noise pollution
- Bad sanitation/smell
- No public restrooms
- Environmental conservation efforts
- Solid waste in waterways
- Passenger and staff safety
- Emergency protocols
- Water quality/toxicity
- Access for those with mobility challenges
- Signage for tourists

Opportunities

- Increasing capacity of public transportation
- Local tourism expansion
- Sustainable khlong transport redevelopment
- Staff training and job enhancement
- Biocultural conservation efforts
- Long term destination sustainability
- Increase awareness of cultural opportunities
- Environmental interpretation
- Cultural interpretation

Threats

- Changing livelihoods
- Unsustainable urban development
- Khlongs being filled in for other infrastructure
- Emissions from riverboats/greenhouse gases
- Climate change/water level rise
- Increase in tariffs for locals
- Possible tourist overcrowding that would impact locals use of transport
- Possible filling-in of remaining khlongs due to modern urban development
- Accidents and illness from contaminated water exposure

Discussion and future research

This chapter explores the intersection between khlong transportation, social capital, and sense of place from a neolocalism lens. Through pre-departure discussions of expectations, semi-structured researcher observations, photography, and a literature review, we conclude that khlong transportation offers a unique experience within Bangkok for both locals and visitors. However, it can be further redeveloped to improve operational conditions, safety, signage, restroom availability, and interpretation for passengers. In addition, we anticipate that with improvements to the express boat conditions and services, there could be an increase in tourist use and better day-to-day transportation to improve locals' lives. If there were to be a focused initiative with consideration of biocultural conservation, as it is relates to redefining a significant element of Bangkokian history and current livelihoods, both locals and tourists could benefit. Khlong transport redevelopment could enhance the tourism experience while benefiting the khlong communities if it is further understood and if initiatives can foster government, private sector, and local community decision-making, management, and jurisdiction. In addition, khlong transport redevelopment could serve as a form of biocultural conservation from a heritage perspective and from a socio-environmental management perspective. The waterways are unique forms of Bangkok heritage and through better management, could also become safer and more enjoyable forms of local and tourist transport. Neolocalism is a way to strengthen insight and communication about, and benefits from, local cultural uniqueness.

Through the redevelopment and conservation of this culturally rich heritage transport, local communities could continue to benefit from efficient transportation. It is recommended that there be a concerted effort to involve multiple stakeholders for future redevelopment of khlong neolocal transport. From previous studies cited in this chapter, this multi-stakeholder involvement could ease potential for conflict of use and increase interest between local communities and visitors/tourists. A focus on improving transport signage could include route information along with cultural and environmental interpretation (along with safety and historical/modern significance of the khlongs).

It has been determined that canals have the potential to develop into a tourism destination as there are usually diverse cultural and historic assets alongside canals. However, a sustainable approach should be taken to develop canal tourism, which requires taking sociocultural, economic, environmental, and climate change concerns into consideration. Additionally, the cooperation of various stakeholders also has been recognised as a key point for the achievement of effective canal management and development. This study explores the participant experience of khlong transportation from the observation of a researcher/tourist perspective and relates to the transportation that local people and tourists experience. As aforementioned, through data analysis and comparative discussions, the researchers generated a SWOT analysis regarding issues related to Saen Saeb khlong redevelopment in Bangkok (see Table 8.1). This can serve to provide insights and suggestions to solve the existing problems and, thereby, improve the current situation for a sustainable path forward.

This analysis suggests that future research should consider measuring benefits and values from both social and environmental dimensions, which focus on quality of life, security, and safety (Lenka & Tiwari, 2015). These challenges and solutions need to balance sustainability with profit-making and management (Fry & Slocum, 2008). By utilising a multi-stakeholder approach focused on issues of community benefit and biocultural conservation, khlong redevelopment could enhance both tangible and intangible assets. In addition, future research will need to integrate issues of climate change impacts as related to waterway management in Bangkok. The urban impacts resulting from climate change, such as increasing temperatures, rising sea levels, as well as extreme weather changes, such as storms, floods, drought, and cyclones (Lukito et al., 2019), need to be considered as they relate to transportation modalities in Bangkok. In addition, if a redevelopment initiative is approached inclusively and effectively, it could also serve as a form of climate change resilience in the modern urban context.

The outcome from this philosophical and empirical engagement with neolocalism from the perspective of transportation is linking the conceptualisations of valuing historic and modern transportation as an underpinning of identity and sense of place and reinforces cultural and social capital. Through the conservation and further sustainable redevelopment (implementing appropriate interpretation and other operational considerations, such as health and safety issues) of this type of historical and modern transport route via the remaining khlongs, both locals and visitors can be exposed to an important keystone of Bangkok's sense of place. Experiencing a khlong in Bangkok provides a unique perspective into the post-modern reality of locals, the infrastructure of an ever-changing megacity, the historical routes of an ancient culture, and the issues that are involved with biocultural conservation in the city today.

Hence, khlong river transport can be viewed and further understood through an alternate lens. If the ideology behind the notion of neolocalism is applied to this context, perhaps it can serve as conduit for both locals and tourists alike to embrace khlong transport as an issue of biocultural conservation. The future of this khlong transport system can be protected, and perhaps the cultural value of the ancient water network can outlast modern urban development. Furthermore, via the applied lens of neolocalism, it is suggested that conservation and clean up initiatives can be furthered by multiple stakeholders, including the local community. Future research is required regarding the connection between biocultural conservation, social capital, transportation, and neolocalism. When a community and visitors are provided information and interpretation of a neolocal attraction, can it serve to further the protection of social capital? Can historical transportation modes and routes be conserved through rebranding and tourism interpretation? What are the risks and, therefore, management approaches to ensure equal access to these neolocal tourism experiences when locals rely on the service for daily life?

References

Barnham, C. (2015) Quantitative and qualitative research, *International Journal of Market Research*, **57**(6), 837-854.

Batra, A. (2014) Floating markets: balancing the needs of visitors as a tourist attraction and locals way of life. A case study of Talingchan Floating Market, Bangkok Thailand, *International Journal of Hospitality & Tourism Systems*, **7**(2), 1-9.

Bangkok Post. (2018) Canals eyed for tourism, play study, 2 February. Available at http://www.bangkokpost.com/thailand/general/1405974/canals-eyed-for-tourism-play-study. (Accessed: 15/05/2018).

Bangkok Post. (2019) Record 38.27m tourists in 2018; 41m expected in 2019, 28 January. Available at https://www.bangkokpost.com/business/1619182/record-38-27m-tourists-in-2018-41m-expected-in-2019. (Accessed: 7/11/2019).

Bangkok Expressway and Metro (BEM). (2018) Available at https://metro.bemplc.co.th/Fare-Calculation. (Accessed: 3/01/2019).

Bangkok Mass Transit Authority (BMTA). (2019) Available at http://www.bmta.co.th/th/services. (Accessed: 21/11/2019).

Bangkok Transportation System (BTS). (n.d.) Available at https://www.bts.co.th/routemap.html. (Accessed: 13/11/2019).

Bodry, C. (2012) Exploring Bangkok's canals, BBC, 29 August. Available at http://www.bbc.com/travel/story/20120827-exploring-bangkoks-canals. (Accessed: 27/08/2018).

Curtis, F. (2003) Eco-localism and sustainability, *Ecological economics*, **46**(1), 83-102.

Duffy, M. & Mair, J. (2018) Engaging the senses to explore community events, *Event Management*, **22**(1), 49-63.

Department of Drainage and Sewerage of Bangkok. (2019). สานํ ักการระบายนํา้, 8 December. Available at http://dds.bangkok.go.th/. (Accessed: 18/11/2019).

Flack, W. (1997) American microbreweries and neolocalism: "Ale-ing" for a sense of place, *Journal of Cultural Geography*, **16**(2), 37–53.

Fry, L.W. & Slocum, J.W., Jr. (2008) Maximizing the triple bottom line through spiritual leadership, *Organizational Dynamics*, **37**(1), 86-96.

Gill, J. & Johnson, P. (2002) *Research methods for managers*, Newbury Park, CA: SAGE.

Giriwati, N., Homma, R. & Iki, K. (2013) Urban tourism: Designing a tourism space in a city context for social sustainability, *The Sustainable City VIII*, **1**, 165-176.

Iamtrakul, P., Srivanit, M. & Klaylee, J. (2017) Resilience in urban transport towards hybrid canal-rail connectivity linking Bangkok's canal networks to mass rapid transit lines, *International Journal of Building, Urban, Interior and Landscape Technology; BUILT*, **10**, 27-42.

Iamtrakul, P., Raungratanaamporn, I. & Klaylee, J. (2018) Contribution on water transportation for resilient and sustainable lowland cities, *Lowland Technology International*, **20**(3), 341-350.

Inland Water Transport Development Company. (2018) Advantages of Inland Water Transport System, IWTDC, Government of The Punjab. Available at http://iwt.punjab.gov.pk/aiwts. (Accessed: 17/08/2019).

Jansen, A. (2015) Positioning and subjectivation in research interviews: Why bother talking to a researcher?, *International Journal of Social Research Methodology*, **18**(1), 27-39.

Jiarakul, D. (2016) Khlong: Bangkok canals revitalization, *Veridian E-Journal, Silpakorn University: (Humanities, Social Sciences and Arts)*, **9**(5), 185-196.

Kaaristo, M. & Rhoden, S. (2017) Everyday life and water tourism mobilities: Mundane aspects of canal travel, *Tourism Geographies*, **19**(1), 78-95.

Khaosod. (2016) คลองแสนแสบมีมาตังแต้ เม่ ือไร, 10 October. Available at https://www.khaosod.co.th/lifestyle/news_44179. (Accessed: 8/08/2018).

Kritiyutanont, K., Nagai, M. & Witayangkurn, A. (2016) Transportation modes detection in Bangkok using GPS logger data and GIS data, *TNI Journal of Engineering and Technology*, **4**(2), 52-60.

Lenka, U. & Tiwari, B. (2016) Achieving triple 'P' bottom line through resonant leadership: an Indian perspective, *International Journal of Productivity and Performance Management*, **65**(5), 694-703.

Lukito, J.A., Susilawati, C. & Goonetilleke, A. (2019) Climate change adaptation in the management of public buildings: An Indonesian context, *International Journal of Building Pathology and Adaptation*, **37**(2), 140-162.

Majumdar, B. (2018) Exploring the local tourism trend, *Multibriefs: Exclusive*, 3 October. Available at http://exclusive.multibriefs.com/content/exploring-the-local-tourism-trend/travel-hospitality-event-management. (Accessed: 20/12/2018).

Marine Department. (n.d.) อัตราโดยสารคลองแสนแสบ ประจาปี 2561. Available at https://www.md.go.th/stat/images/pdf_report_stat/2561/M_Saen_Saep_61.pdf. (Accessed: 3/11/2019).

Organisation for Economic Cooperation and Development. (2018) *Tourism Trends and Policies*. Paris: OECD Publishing. Available at http://dx.doi.org/10.1787/tour-2018-en. (Accessed: 7/11/2019).

Parnwell, M.J. (2007) Neolocalism and renascent social capital in northeast Thailand, *Environment and Planning D: Society & Space*, **25**(6), 990–1014.

Phaithayawat, S. (2015) Co-existence of Thai Muslim people and other in an ancient community located in the heart of Bangkok: The case study of Petchaburi 7 community, *World Academy of Science, Engineering and Technology, International Journal of Social, Behavioral, Educational, Economic, Business and Industrial Engineering*, **9**(3), 827-830.

Prakraiwan, T. (2016) Cultural tourism route of Thai way of life along Bangkok canal. *NAJUA: Architecture, Design and Built Environment*, **31**, 135-157.

Revfine. (n.d.) Tourism trends: 13 opportunities for the tourism industry. Optimising Revenue. Available at https://www.revfine.com/tourism-trends/. (Accessed: 1/11/2019).

Saunders, M.N. (2011) *Research Methods for Business Students*, 5th edn., Pearson Education India.

Schnell, S. (2013a) Deliberate identities: Becoming local in America in a global age, *Journal of Cultural Geography*, **30**(1), 55–89.

Schnell, S. (2013b) Food miles, local eating, and community supported agriculture: putting local food in its place, *Agriculture and Human Values*, **30**(4), 615–628.

Shipley, R. (1999) Making a silk purse from a sow's ear: lessons from the failure to develop the Welland Canal as a tourism resource, *Environments*, **27**(2), 25.

Shortridge, J.R. (1996) Keeping tabs on Kansas: Reflections on regionally based fieldstudy, *Journal of Cultural Geography*, **16**(1), 5–16.

Suwandee, S., Anupunpisit, V., Ratanamaneichat, C., Sukcharoen, N. & Boonpen, P. (2013) Quality of life and environment of communities along Saen Saeb

Canal: A survey foundation of the physical and the current situation (Phase I), *Procedia - Social and Behavioral Sciences,* **88**, 205–211.

Tang, L. & Jang, S. (2010) The evolution from transportation to tourism: The case of the New York canal system, *Tourism Geographies,* **12**(3), 435-459.

Transit Bangkok. (n.d.) Complete guide to public transportation in Bangkok. Available at https://www.transitbangkok.com/th/khlong_boats.html. (Accessed: 12/09/2019).

UNWTO. (2011) Tourism Towards 2030: Global Overview, United Nations World Tourism Organization Madrid, 10 October. Available at https://www.globalwellnesssummit.com/wp-content/uploads/Industry-Research/Global/2011_UNWTO_Tourism_Towards_2030.pdf. (Accessed: 15/08/2018).

Ujang, N. & Zakariya, K. (2015) The notion of place, place meaning and identity in urban regeneration. *Procedia Social and Behavioral Sciences,* **170**, 709-717.

Wang, Y. (2012) Management of the Grand Canal and its bid as a world cultural heritage site, *Frontiers of Architectural Research,* **1**(1), 34-39.

Wang, S. & Xu, H. (2015) Influence of place-based senses of distinctiveness, continuity self-esteem, and self-efficacy on residents' attitudes toward tourism, *Tourism Management,* **47**, 241-250.

World Population Review. (2019) Bangkok Population. Available at http://worldpopulationreview.com/world-cities/bangkok-population/. (Accessed: 20/09/2019).

Acknowledgements: The authors would like to thank Dr. Adarsh Batra and Dr. Jason Swanson for their support during the doctoral course.

9 Neolocalism and Social Sustainability: The Case of Öland's Harvest Festival, Sweden

Guanhua Peng, Solène Prince and Marianna Strzelecka

Abstract

Food festivals are common happenings in rural communities. Arguably, these events are a form of neolocalism, where the efforts of local stakeholders promote a sense of place based on attributes of their community. In this chapter, we propose that food events foster residents' sense of place, and thus contribute to rural social sustainability. We explore the case of the annual harvest festival on the island of Öland, Sweden (Ölands Skördefest). This festival seeks the promotion of local cuisine, innovative culinary products and rural culture. We argue that it even fosters feelings of belonging amongst community members by generating an atmosphere of cooperation and social participation. Participation in Ölands Skördefest can strengthen new residents' feeling of belonging to a special place. Moreover, the event facilitates stable business networks on Öland as different stakeholders work together to organize an attractive event. The work of these individuals reinforces a sense of belonging to a special place during the organization of the event. Memories of the festival amongst community-members also cultivate the continuity of positive social relationships on the island.

Introduction

Food festivals are common events taking place in rural communities where local produce, traditions, and modes of production are displayed and celebrated for community members and visitors alike to enjoy. Arguably, food events are a form of neolocalism, where the efforts of local stakeholders promote a sense of place based on attributes of their community (Holtkamp et al., 2016; Schnell, 2013). Local foods have gained prominence in the past years due to growing popular concerns over sustainable food production and feelings of alienation from these modes of production (Mason & Paggiaro, 2009; Mykletun & Gyimóthy, 2010; Sims, 2009). In many rural communities, local foods and food production are significant markers of place-identity (Caplan, 1997; Cook & Crang, 1996). As Timothy and Ron (2013) argue, local gastronomies have developed in different ways in different places, making foods a significant component of place uniqueness, and subsequently, of the sense of place it fosters among community members. We propose that food events, as neolocalism, foster residents' sense of place, and thus contribute to the social sustainability of rural communities.

To build the aforementioned argument, we explore the case of the annual harvest festival on the island of Öland, Sweden (Ölands Skördefest). This festival not only revives old farming traditions, it also promotes and supports local cuisine, innovative culinary products, and rural culture (https://www.skordefest.nu/english/). Besides extending Öland's summer tourist season, the local businesses co-organising the festival directly state that they seek to achieve a sense of community across Öland with this particular event. That is, they seek to foster a sense of place based on foods and crafts related to the attributes of local villages. Ölands Skördefest is, thus, a prime example to study the emergence of neolocalism in tourist destinations through community-based events. The results of this study come from in-depth interviews conducted with different participants involved in Ölands Skördefest, including event organisers, tourism officers, hospitality workers, farmers, and restaurant owners.

We identified two themes that outline the formation of a sense of place through a neolocal event and their relationship to community social sustainability. First, the unique character of Ölands Skördefest fosters feelings of belonging amongst community members in the light of, not only the distinct products and produce put on offer during the event, but also of the atmosphere of cooperation and social participation it generates. Moreover, participation in Ölands Skördefest can strengthen new residents' feelings of belonging to a place with a special culture. Second, the event facilitates stable business networks on Öland as different stakeholders work together to organise a cohesive and attractive food festival that will benefit all players involved. The hard work of these individuals, based in solidarity and pride, reinforces a sense of belonging to a special place during the organisation of the event. Moreover, memories of the festival amongst community members cultivate the continuity of positive social relationships on the island. The emergence of neolocalism through food events is, thus, significant to the social sustainability of this rural community.

Identity and place in social sustainability

For a long time, the social components of sustainability were mostly disregarded in discourses of sustainable development (Dempsey et al., 2011; Vallance et al., 2011), which is why the term 'social sustainability' emerged to resolve the imbalance and provide a lens through which to focus on social goals. Social sustainability embraces social behaviours, cultural values, lifestyles, and even modes of production and technological advancements, as well as other social constructions (Pol, 2002). Social sustainability, thus, usually relates to topics as diverse as health and social equity, human rights, labour rights, practices, and decent working conditions, social responsibility and justice, community development and wellbeing, product responsibility, community resilience, and cultural competence (Balaman, 2018). Correspondingly, Choi and Turk (2011) argue that scholars, policy-makers, and practitioners should focus more on social and community realms when discussing sustainability issues.

Theorists have offered a plethora of approaches in an effort to gain greater insight into the diverse nature of community processes, often framing the community in terms of place, interaction, and affective bonds, all of which are manifestations of social relations. Following Wilkinson (1991) and Theodori (2005), we embrace an interactional approach to study rural communities and social sustainability. We view a rural community as a dynamic interactional phenomenon, which emerges from interactions between local interest groups involved in matters such as education, environmental protection, governance, and tourism, within the local social context (Theodori, 2005; Wilkinson, 1991). Participants in these different groups unite in 'a process of interaction through time, with direction toward some more or less distinctive outcome' (Wilkinson, 1972: 317). Local actions, such as local events and festivals, tend to facilitate the emergence of an interactional community. Consequently, through an interactional approach, we can conceptualise community development as a 'process of community change, [in which] ecological, organisational, situational, and other forces converge to structure and alter the relationship among people in a local setting' (Wilkinson, 1991: 92). Here, organised action improves coordination among local interest groups (Theodori, 2005). Local identities, culture, and social cohesiveness are all key components of interactional community development where local stakeholders act together and connect with others to realise their common interest.

In the context of interactional community development, community social sustainability requires that local stakeholders are active agents participating in their own development strategies, whereby they are able to mobilise local and extra-local resources in order to meet their own needs, through their own decision-making processes (Stone, 1989). Fulfilling such ambitious community sustainability goals is founded on a particular way of organising social relationships, a general quality of sociability, a mutual regard, and a summons to undertake joint social actions (Day, 2006: 6). The elements of community fabric that enable and encourage local actors to engage in their communities are foundational to its social sustainability. Ultimately, community social sustainability implies that communities are:

> '[...] places where people want to live and work, now and in the future. They meet the diverse needs of existing and future residents, are sensitive to their environment, and contribute to a high quality of life. They are safe and inclusive, well planned, built and run, and offer equality of opportunity and good services for all' (ODPM, 2006: 12).

In this regard, community social sustainability relates to matters such as participation, stability, pride, and security in local development (Dempsey et al., 2011; Vallance et al., 2011; Strzelecka, 2012). This makes place-identity formation a core element in the process towards community social sustainability (Hay, 2010; Pol, 2002; Pol et al., 2002).

Place is a set of spaces transformed into a meaningful location through peoples' experiences and ideas (Leonard, 2014). Relph (1976) distinguished three components of place: physical setting, activities, and meanings, with meanings arguably being the most difficult component to grasp. Place meaning is about how place matters in individuals' every day practices (Agnew, 2011; Gustafson, 2001, 2014). People self-define through places. They make places the centre of symbolic meanings, transforming spaces into symbolic landscapes (Brown & Perkins, 1992; Greider & Garkovich, 1994). These meanings are then reinforced through local practices and rituals (Cresswell, 1996). In the process of meaning creation and co-creation, people develop a place-identity (Williams et al., 1992), which arises from particular values, attitudes, and beliefs about the physical world, as well as from direct experiences with the environment (Proshansky et al., 1983).

A positive sense of place, and its various attributes, translates to a positive sense of identification to the place, which is integral to people's enjoyment of their living space (Nash & Christie, 2003). For Tuan (1974, 1977), one's identity stems a great deal from the association one feels towards a place. The natural features of place generate awe, and the modifications wrought by successive generations of humans generate affection (Tuan, 1974). A sense of place produces collective meaning and identity, from which particular social behaviours and decisions will emerge (Stokowski, 1991). This has significant implications for sustainability; as local individuals cherish the visual and emotional cues that shape their place into a unique socio-cultural realm, they will be more likely to behave in a way to preserve those attributes and support each other in this process (Vaske & Kobrin, 2001; Williams & Vaske, 2003). The sense of place that stakeholders have often leads them to work on the protection and, even, the revitalisation of its distinctive character (Norton & Hannon, 1997; Strzelecka et al., 2011). As Manzo and Perkins (2006: 347) argue, people are motivated to seek, stay in, protect, and improve places that are meaningful to them. It is arguably relevant to study the processes that generate a positive place-identity amongst community members in order to effectively define and promote social sustainability.

Food events as neolocalism

Local foods have gained prominence in the past years due to growing popular concerns over sustainable food production and feelings of alienation from these modes of production. Subsequently, an increasing number of tourists are seeking to consume local foods and drinks to feel a more authentic connection to the place and people they encounter during their travels (Mason & Paggiaro, 2009; Mykletun & Gyimóthy, 2010; Sims, 2009). Localised food production is part of a global movement of reconnecting people with the distinct qualities of place in a climate of globalisation and a lost sense of identity, which geographer Shortridge (1996) dubbed neolocalism. Neolocalism defines the conscious efforts of local stakeholders to promote a sense of place based on attributes of their community (Holtkamp et al., 2016; Schnell, 2013). Neolocalism aims to restore the notion of, and the focus on, local communities. Food is a significant marker of local community distinctiveness and, thus, has the potential to contribute to place-identity formation among community members (Caplan, 1997; Cook & Crang, 1996). As Timothy and Ron (2013) argue, local gastronomies have developed in different ways in different places, making food a significant component of place uniqueness, and, subsequently, of the sense of place it fosters among community members. The production of local foods has a greater purpose than feeding people and generating economic sustainability; it creates and brings back ways of life grounded in place-based attributes and symbols. For instance, Mykletun and Gyimóthy (2010) outline how the old culinary tradition of the sheep-head meal resurfaced in the 1990s in Norway to become a significant place-based experience for locals and visitors.

Food has become a major component in the promotion of many tourism destinations around the world (Getz & Robinson, 2014; Hjalager & Corigliano, 2000). Studies have linked different forms of tourism, such as rural tourism, agritourism, and cultural tourism, to the sustainability of all sorts of food production (Murray & Kline, 2015; Sims, 2009). Che et al. (2005) explain that the development of local food and agricultural produce for rural tourism purposes helps retain and revive local traditions in the face of pressures from global trends of commercialisation and standardisation. Moreover, Getz and Robinson (2014) contend that food and agricultural events are especially important for rural communities, as their event-goers greatly value such genuine food experiences.

Arguably, food events reflect the resurgence of place through conscious local efforts implied in neolocalism, as they are prime venues to promote and strengthen rural communities and their cultures. Mykletun and Gyimóthy (2010) argue that strong networks and entrepreneurial activities are important for the renaissance of culinary traditions. However, researchers tend to pay more attention to the environmental and economic sustainability of events, such as harvests, farmers' markets, and agricultural and food fairs. Food events and their relation to social sustainability, such as community cohesion, pride and participation, is still a relatively unexplored topic in current tourism research.

Generally, researchers acknowledge that community events can foster social sustainability and regard them as a means to motivate local participation, increase social cohesion, and create an effective place-brand (see for example: Barrera-Fernández & Hernández-Escampa, 2017; Davis, 2016; Moscardo, 2007; Richards, 2017). Moreover, local festivals can foster positive self-identification at the community level when they aim to promote and preserve local culture and tradition (De Bres & Davis, 2001). Jaeger and Mykletun (2013) contend that events can promote place-identity through their celebration and display of local culture in the face of rapid demographic and socio-cultural change, and consequently fulfil a need to belong to a community amongst new and old members. Similarly, Derrett (2003) suggests that community events strengthen the values, interests, and aspirations of residents who use the occasion to display their special place and share stories of it with visitors, friends, and relatives. Coghlan et al. (2017) outline that organisers should see the added benefits of strategically planning events for place-making purposes, going beyond economic benefits and improving local liveability. In light of the positive effects that local events can have on rural communities, it is interesting and important to better understand the role of food events as acts of neolocalism, and their contribution to the social sustainability of rural places.

Case study: Ölands Skördefest

Located in the Baltic Sea, Öland is the second largest island in Sweden (1347 km^2). It is renowned for its idyllic countryside and sandy beaches and connects to the Hanseatic city of Kalmar through a six-kilometre long bridge. Öland's inhabitation dates as far back as the Bronze Age and is, thus, an island rich in history. Nowadays, Öland is home to nearly

26,000 inhabitants who work either on the island or in Kalmar. The island retains its idyllic charm through its cultural landscape of small fishing villages and family farms. The small family farms produce all sorts of goods, including honey, meat and dairy products, and harvest crops such as beans, rapeseed, vegetables, and strawberries, which they sell locally and even around and beyond the region in some cases (see Images 9.1 and 9.2).

Images 9.1 and 9.2: Many different goods are farmed, transformed, consumed and sold on Öland. These two pictures show rapeseed oil and honey made on Öland. Photos by Marianna Strzelecka, 2019.

Over the years, Ölands Skördefest has become the most popular harvesting event of its kind in Sweden, attracting high numbers of visitors since its creation in 1997. Some 500 local businesses participate in Ölands Skördefest. The thousands of different activities taking place on the island during the event involve various family farms that sell their products and handicrafts, and restaurants that promote the local cuisine such as the famous *kroppkakor*. Kroppkakor is a traditional dish from the area of Sweden to which Öland belongs. It is a form of dumpling filled with pork and onions, usually served with butter and lingonberry jam.

Ölands Skördefest runs through a joint incorporated company and belongs to Skördefestens Vänner Ideell Förening – a non-profit association formed of local individuals and companies who aim to promote the popularity and long-term development of Ölands Skördefest. Ölands Skördefest is a celebration of the ancient tradition Michaelmas, which centres on the harvest of crops before winter. More specifically, Michaelmas is an ancient Catholic tradition, dating back to the rise of Christianity in Rome, that includes feasting and fasting in honour of Saint Michael and other important angels. In the northern hemisphere, it is associated with the arrival of autumn and the shortening of daytime. It is still celebrated in Western countries, such as Britain and Sweden, and in various places in North America. Over time, the celebration has lost most of its religious meaning to become a secular event during which farmers and business owners sell their goods, and community members have a good time.

The organisers of Ölands Skördefest believe that Öland's special character and tradition is a strong brand that can support success on the Swedish and international marketplace. Ölands Skördefest functions primarily to prolong the island's summer tourist season and support its fine cuisine, its innovative culinary products, and its rich cultural heritage. Ölands Skördefest exists as a means to achieve a sense of togetherness amongst the diverse communities of the island through its varied activities. Importantly, the event should direct attention towards, and increase the accessibility to, local produce, handicrafts, and culinary innovations and create better business opportunities for local stakeholders.

Methods

Since the study relates to experiences, social dynamics, and human senses, a qualitative approach was deemed essential to explore this case. During fieldwork in 2018, in-depth interviews were conducted with ten different stakeholders involved in the promotion, development and delivery of Ölands Skördefest, including event organisers, tourism officers, hospitality workers, local farmers, and restaurant owners. The interviews lasted approximatively one hour each and were all recorded and then transcribed for analytical purposes. The respondents talked about how their daily lives, careers, and/or business were shaped by Ölands Skördefest. Respondents also reported on their experiences of the festival, as well as their feelings towards Öland as a place of residency. Our results point at two themes that outline the formation of a sense of place and the social sustainability of the Öland community, presented below as: community belonging through participation in a unique event; and working together in solidarity for pride and continuity.

Community belonging through participation in a unique event

Several study-participants positioned Öland's agricultural products and cuisine as distinctive of the island and, thus, as fundamental to the uniqueness of Ölands Skördefest. The small-scale food production is special because of its particular history and cultural significance that tie it to Öland, and consequently, position the island as a unique place in the eyes of its inhabitants. Such place-based narratives over food production support the idea that food is a significant component of the uniqueness of a place (Timothy & Ron, 2013). Ölands Skördefest is a place to celebrate these local traditions but also to reinforce the distinctiveness of the place in the eyes of customers. Neolocalism is apparent in this quote from a study-participant about the special Öland brown bean (see Image 9.3) and the meaning it has for local people and visitors:

> 'People are coming here for our special organic food, which is produced and cooked on Öland. For example, Öland is the biggest bean production area of Sweden with a long history, and we are trying to use our local brown bean to introduce Öland, also our ideas of spirit and passion to our customers.'

Image 9.3: This photo shows different sorts of peas and beans from Öland sold in bulk at a local boutique, including the special local brown bean (*Öländska bruna bönor*). Photo by Marianna Strzelecka, 2019.

As Derrett (2003) argues, the unique values, interests, and aspirations of local stakeholders find expression in community-based events, which bolsters an awareness amongst these individuals of the special character of their community. There is a conviction amongst the stakeholders on Öland that the uniqueness of their food products and cuisine attracts visitors to the island, but also that the island in itself and the festival are attractive to visitors because of their unique character. Scholars have found that a positive sense of place distinctiveness contributes to community pride (Boley et al., 2018). In this case, the festival reinforces a positive sense of place distinctiveness crucial to community pride. The following two quotes highlight how the harvest festival is part of a bigger distinct culture for its people:

> 'Some tourists from Stockholm talk with me and they say that they really enjoy their holiday on Öland. They come here for not only the festival, but also the air and sunlight of Öland, which is quite different from the northern area [of Sweden], and they think they are more comfortable.'

> 'I mean there are many other harvest festivals in Sweden, but the harvest festival on Öland is unique because it is on the island. You can't find similar harvest festival experiences outside Öland, that is the reason why [the tourists] keep coming back.'

The study participants also consider the social environment that the preparation and celebration of Ölands Skördefest fosters as unique. From the perspective of tourism stakeholders, Ölands Skördefest nurtures communal harmony as the residents and local entrepreneurs come together to cooperate in the delivery of a high-quality event. As established, the sense stakeholders have of their place often leads them to work on the revitalisation of its distinctive character (Norton & Hannon, 1997; Strzelecka et al., 2011). In this case, the revitalisation of the place-based character occurs through the organisation of a festival that sells and celebrates cherished local foods and promotes culinary innovations. To these stakeholders, this type of positive community relationships is very special and is not evident in the organisation of other harvest festivals in Sweden.

The sense of place promoted in neolocal efforts manifest not only in the special character of cultural products, but also in the special character of people's interactions during these efforts. This result suggests that the event facilitates interaction and cooperation among different actors of the Öland community, thus facilitating community development. The two quotes below outline the perceived special character of the event because of the social relations and sense of togetherness is generates between different members of the community:

> 'The whole island gets together, works together, the small businesses, as well as the bigger businesses. Maybe a business just operates for fun, it is not professional, but it shows their passion.

> You want to show it to everyone. This event welcomes everyone on the island to come to join. That makes the festival unique.'
>
> 'When the event takes place, Öland transforms into something quite different from what other events usually offer. Much of this is because of the cooperation between stakeholders and companies cooperating to sell their [food] products.'

Ultimately, the positive feelings of displaying and producing something unique fosters not only cooperation but also a sense of belonging to a special place amongst the local stakeholders, similarly as other researchers have found through their studies of community events (De Bres & Davies, 2001; Derrett, 2003; Jaeger & Mykletun, 2013).

Moreover, Ölands Skördefest seems to enhance newcomers' sense of belonging to Öland as a place of residency because the event happens on the whole island. As Jaeger and Mykletun (2013) demonstrate, the inclusion of newcomers in local events strengthens their sense of belonging to a community. Residents on Öland can get involved and participate in an event that connects them with different, smaller communities around the island and makes them feel accepted within a larger insular community. These newcomers' enjoyment of their new place of residency, thus, relates to a positive identification to the place and its people, as Nash and Christie (2003) contend. The two quotes below are from newcomers who consider the process of organising Ölands Skördefest as an opportunity to strengthen their relationship with their fellow community-members. The quotes demonstrate how an event can foster a sense of community belonging among its participants:

> 'This event definitely changes my relationship with the community. Actually, I don't really have business relationships with other local people, but this event creates friendships. That brings us a lot of joy. Whatever the weather is during that period, my friends and I are happy at the festival. '
>
> 'Personally, I am so happy to join this big event. I really enjoy the feeling of working with other people during the festival. That makes me feel I am a part of this big thing.'

As an act of neolocalism, the harvest festival fosters social sustainability within the Öland community because it creates a venue for multiple stakeholders, including new residents, to feel a sense of belonging as they interact together and engage in local consumption and event participation, or in event organisation and promotion.

Working together in solidarity for pride and continuity

One cannot overlook that Ölands Skördefest enhances the participation of small businesses in the island's economy. This participation is crucial since Öland faces many of the challenges typical of rural destinations in Nordic countries, such as seasonality and declining levels of population. Many respondents agree that the festival is integral to the economic sustainability of small agricultural producers and restaurants on the island. Though July is the most intense month of the tourist season, business owners make more profit during the weekend of Ölands Skördefest than in July, according to a study participant. This supports the claim from Hall and Sharples (2008) that food events in rural community are economically significant. In this case, it extends the tourist season during harvest time in the autumn and puts the spotlight directly on these small companies and family farms seeking extra revenue (see Image 9.4).

Image 9.4: Many local boutiques and farmers benefit from the prolonged tourist season that the harvest festival promotes. This is the case for this boutique in Algutsrum, which sells its own organic fruits and vegetables,and other local produce and products such as milk, eggs, meat, jam, and honey. Photo by Marianna Strzelecka, 2019.

As Mykletun and Gyimóthy (2010) established, the revival of local culinary traditions depends on the entrepreneurial activities and networking of stakeholders embedded in a place. Similarly, Getz and Andersson (2010) contend that events can stimulate positive and stable business networks between small and big businesses. However, these networks are not only important for the smaller players' economic sustainability, as through these networks, they can extend their sales and marketing opportunities. Business connections are also important for social sustainability, as they stimulate interaction and cooperation amongst various stakeholders. This way, these active networks strengthen community pride and cohesiveness. The quote below illustrates the social contribution of business networks in event organisation in bringing community members together to achieve local goals:

> 'The event has brought a lot to Öland over the years. Smaller companies that don't have an opportunity to have their business open all year around are given the chance to be heard of and to offer their products to people all over Sweden and the world. This has brought the community closer together.'

Huang and Stewart (1996) argue that cooperation built around common goals of survival reinforces community solidarity. In this case, the local stakeholders rally together in their neolocal effort to promote creativity, products, and the produce of Öland, and find support in each other in this endeavour. Effective cooperation then leads to feelings of pride that they do this successfully together, as the quote below illustrates:

> 'Since I am a co-organiser of this festival, [I work with it] for the whole year, I get to know everyone around the island, from the north to the south. These people think this festival can bring the community together, and everyone is proud to be a part of the Öland harvest festival. This is our achievement.'

Besides generating positive feelings of being able to cooperate, the successful organisation of the event generates positive feedback from tourists, which further reinforces a sense of pride amongst community-members, as these two quotes demonstrates:

> 'Because I am a part of [the harvest festival on Öland], I will feel proud when I hear people making positive comments about it.'

> 'For me, if we can create a good experience [for tourists], which encourages people to come back, it provides positive feedback to me and it makes me proud.'

Economic stability and business profit are not the only positive outcomes of local networking during food events, positive feelings of accomplishment and strengthened social relationships are arguably important outcomes of neolocalism. As established, sustainability cannot be dissociated from social interactions, as they enable and guide participation in community affairs (Choi & Turk, 2011; Pol, 2002). The notion of achieving something meaningful together spurs narratives of solidarity amongst community members who work hard in order to offer a unique experience to event goers. In these quotes, it is shown that the hard work of organising the event brings positive feelings to local stakeholders:

> 'Even though I see the same scene every year, it still makes it unforgettable...the busiest weekend in the whole year! But it feels good, everyone is working hard and we all do our best to create good experiences for our customers.'

> 'The opening of the festival takes place in different places of the island every year. Last year was very special, it was in a little village in Vickleby, everyone was so happy and got involved in the activities. Everyone wanted it to be the best one and that is impressive.'

Besides reminiscing about their hard work, study-participants also spoke of their good experiences with friends and family during Ölands Skördefest. The festival provides the environment and atmosphere for local stakeholders to enjoy and maintain existing social relationships, as they come together to display something special to event-goers (see Image 9.6). The quote below illustrates how family festival traditions, in this case serving customers in a costume, foster positive social relations, as well as entertains customers:

> 'I have unforgettable memories for every year! Do you know the tradition of having a costume for Halloween? We dress in special costumes in the shop for the harvest festival every year: my wife and me, my sister is here, my father is here... we have different themes every year. We create a good memory with my family during the festival. We have done this tradition for seven years. People come here during the festival to see how we dress this year.'

The fondness that respondents express for Ölands Skördefest through their memories translates into social sustainability, as it cultivates the continuity of social relationships on the island. The event creates what Dixon et al. (2014) conceptualise as a stable relationship between place and people over time. Neolocalism is thus a form of social sustainability, as it strengthens the cultural narratives and social ties that uphold a community's fabric through time. Memories of the event, as a dynamic social space, reinforce a sense of belonging to a distinct community. Importantly, Azara et al. (2018) argue that the social continuity created by events can enhance tourism development and maintain the uniqueness of the community.

Image 9.5: This picture shows the harbor in Byxelkrok, a small village located at the top-end of Öland. Some of the special activities in this village during the harvest festival include feasting on crab and enjoying the atmosphere at the harbour. Photo used with permission from Öland's Tourism Office.

Conclusion

We have identified Ölands Skördefest as an act of neolocalism through its promotion and support of local food production and local lifestyles in Öland, Sweden. While the unique place-identity that the festival diffuses boosts the revenue of local businesses by prolonging the tourist season on the island, the contributions this neolocal effort makes to local social sustainability are also noteworthy. Our case event shows that local foods are good markers of place distinctiveness and are, as such, major components of identity formation for rural communities (Caplan, 1997; Cook & Crang, 1996; Timothy & Ron, 2013). This is the case for the rural communities of Öland where the harvest of agricultural produce and the production and consumption of local foods make up both cultural ways of life and the insular landscapes. We positioned these foods and their celebration through Ölands Skördefest as central to the generation of a positive sense of place based in community belonging, pride, enjoyment, solidarity, and continuity. We found it crucial to adopt an interactional perspective on community sustainability, as Wilkinson (1991) popularised in the context of rural farming communities. As the different stakeholders on Öland cooperate to organise an attractive and unique festival, they interact to reach a common goal – this is exactly how communities form according to interactional approaches. An interactional approach, thus, alerts us to the social processes and dynamics that bind the community together, generating its social sustainability.

There are many studies linking community events to aspects of social sustainability, such as belonging, inclusivity, pride, and identity formation (for instance, Derrett, 2003; De Bres & Davis, 2001; Jaeger & Mykletun, 2013). Food events show similar dynamics and outcomes as other community events since they celebrate local cultures and display associated lifestyles to the wider public. Local foods and food events are growing in popularity, as the work of scholars, such as Getz and Robinson (2014) and Mykletun and Gyimóthy (2010) show, and, therefore, there is good reason to study their meaning for rural communities and their social sustainability. However, the foods in themselves are not the only thing unique and special that is put on display during such events, as the case of Öland demonstrates. As we found, the social interactions necessary to the organisation of Ölands Skördefest are a point of distinctiveness according to community members. The process of organising Ölands Skördefest fosters communal solidarity as the residents and local businesses cooperate to deliver a memorable experience to visitors that exhibits place-based elements. The community members derive a lot of pride from their hard work, which fosters a positive sense of belonging. Through their hard work, the organising stakeholders also form positive memories of their social interactions that cultivates a sense of continuity on the island. Neolocalism, thus, contributes to social sustainability precisely because the promotion of unique and cherished local attributes strengthens cooperation between different community stakeholders. Ultimately, neolocal efforts, such as those found in Ölands Skördefest, develop the community fabric by fostering a sense of solidarity and togetherness among event stakeholders, newcomers, and participants.

References

Agnew, J. (2011) Space and place, in J. Agnew & D. Livingstone (eds.), *Handbook of Geographical Knowledge*, London: Sage Publications, pp. 316–330.

Azara, I., Wiltshier, P. & Greatorex, J. (2018) Against all odds: Embedding new knowledge for event continuity and community well-being, *Event Management*, **49**(1), 97-112.

Balaman, S.Y. (2018) *Decision-Making for Biomass-Based Production Chains: The Basic Concepts and Methodologies*, London: Elsevier Academic Press.

Barrera-Fernández, D. & Hernández-Escampa, M. (2017) Events and placemaking: The case of the Festival Internacional Cervantino in Guanajuato, Mexico, *International Journal of Event and Festival Management*, **8**(1), 24-38.

Boley, B., Strzelecka, M. & Watson, A. (2018) Place distinctiveness, psychological empowerment, and support for tourism, *Annals of Tourism Research*, **70**, 137-139.

Brown, B.B. & Perkins, D.D. (1992) Disruptions in Place-attachment, in I. Altman & S.M. Low (eds.), *Place-attachment, Human Behavior and Environment*, Boston, MA: Springer, pp. 279-304.

Caplan, P. (1997) *Food, Health and Identity*, London: Routledge.

Che, D., Veeck, A. & Veeck, G. (2005) Sustaining production and strengthening the agritourism product: Linkages among Michigan agritourism destinations, *Agriculture and Human Values*, **22**(2), 225-234.

Choi, H.C. & Turk, E.S. (2011) Sustainability indicators for managing community tourism, in M. Budruk & R. Phillips (eds.), *Quality-of-Life Community Indicators for Parks, Recreation and Tourism Management*, Dordrecht, Netherlands: Springer, pp. 115-140.

Coghlan, A., Sparks, B., Liu, W. & Winlaw, M. (2017) Reconnecting with place through events: Collaborating with precinct managers in the placemaking agenda, *International Journal of Event and Festival Management*, **8**(1), 66-83.

Cook, I. & Crang, P. (1996) The world on a plate: Culinary culture, displacement and geographical knowledges, *Journal of Material Culture*, **1**(2), 131–153.

Cresswell, T. (1996) *In Place/Out of Place: Geography, Ideology, and Transgression*, Minneapolis, MN: University of Minnesota Press.

Davis, A. (2016) Experiential places or places of experience? Place-identity and place-attachment as mechanisms for creating festival environment, *Tourism Management*, **55**, 49-61.

Day, G. (2006) *Community and Everyday Life*, London: Routledge.

De Bres, K. & Davis, J. (2001) Celebrating group and place-identity: A case study of a new regional festival, *Tourism Geographies*, **3**(3), 326-337.

Dempsey, N., Bramley, G., Power, S. & Brown, C. (2011) The social dimension of sustainable development: Defining urban social sustainability, *Sustainable Development*, **19**, 289-300.

Derrett, R. (2003) Making sense of how festivals demonstrate a community's sense of place. *Event Management*, **8**(1), 49-58.

Dixon, J., Durrheim, K. & Di Masso, A. (2014) Places, identities and geopolitical change: Exploring the strengths and limits of Identity Process Theory, in R. Jaspal & G.M. Breakwell (eds.), *Identity Process Theory: Identity, Social Action and Social Change*, Cambridge: Cambridge University Press, pp. 270-294.

Getz, D. & Andersson, T. (2010) Festival stakeholders: Exploring relationships and dependency through a four-country comparison, *Journal of Hospitality and Tourism Research*, **34**(4), 531-556.

Getz, D. & Robinson, R.N. (2014) Foodies and food events, *Scandinavian Journal of Hospitality and Tourism*, **14**(3), 315-330.

Greider, T. & Garkovich, L. (1994) Landscapes: The social construction of nature and the environment, *Rural Sociology*, **59**(1), 1–24.

Gustafson, P.E.R. (2001) Meanings of place: Everyday experience and theoretical conceptualisations, *Journal of Environmental Psychology*, **21**(1), 5–16.

Gustafson, P.E.R. (2014) Place-attachment in an age of mobility, in L.C. Manzo & P. Devine-Wright (eds.), *Place-attachment: Advances in Theory, Methods and Application*, London: Routledge, pp. 37-48.

Hall, C.M. & Sharples, E. (2008) *Food and Wine Festivals and Events around the World: Development, Management and Markets*, Oxford: Butterworth-Heinemann.

Hay, R. (2010) The relevance of ecocentrism, personal development and transformational leadership to sustainability and identity, *Sustainable Development*, **18**(3), 163-171.

Hjalager, A.M. & Corigliano, M.A. (2000) Food for tourists—determinants of an image, *International Journal of Tourism Research*, **2**(4), 281-293.

Holtkamp, C., Shelton, T., Daly, G., Hiner, C.C. & Hagelman III, R.R. (2016) Assessing neolocalism in microbreweries, *Papers in Applied Geography*, **2**(1), 66-78.

Huang, Y.H. & Stewart, W.P. (1996) Rural tourism development: Shifting basis of community solidarity, *Journal of Travel Research*, **34**(4), 26-31.

Jaeger, K. & Mykletun, R.J. (2013) Festivals, identities and belonging, *Event Management*, **17**(3), 213-226.

Leonard, S.P. (2014) The need to 'belong': Social connectedness and spatial attachment in Polar Eskimo settlements, *Polar Record*, **50**(2), 138-146.

Manzo, L. & Perkins, D. (2006) Finding common ground: The importance of place-attachment to community participation and planning, *Journal of Planning Literature*, **20**, 335-350.

Mason, M. & Paggiaro, A. (2009) Celebrating local products: The role of food events, *Journal of Foodservice Business Research*, **12**, 364–383.

Moscardo, G. (2007) Analyzing the role of festivals and events in regional development, *Event Management*, **11**(1-2), 23-32.

Murray, A. & Kline, C. (2015) Rural tourism and the craft beer experience: Factors influencing brand loyalty in rural North Carolina, USA, *Journal of Sustainable Tourism*, **23**(8-9), 1198-1216.

Mykletun, R.J. & Gyimóthy, S. (2010) Beyond the renaissance of the traditional Voss sheep's-head meal: Tradition, culinary art, scariness and entrepreneurship, *Tourism Management*, **31**(3), 434–446.

Nash, V. & Christie, I. (2003) *Making Sense of Community*, London: Institute for Public Policy Research.

Norton, B.G. & Hannon, B. (1997) Environmental values: A placed-based approach, *Environmental Ethics*, **19**(3), 227-246.

Office of the Deputy Prime Minister (ODPM). (2006) *EU Ministerial Informal on Sustainable Communities Policy Papers*, London: UK Presidency.

Pol, E. (2002) The theoretical background of the city-identity-sustainability network, *Environment and Behavior*, **34**(1), 8-25.

Pol, E., Moreno, E., Guàrdia, J. & Iniguez, L. (2002) Identity, quality of life, and sustainability in an urban suburb of Barcelona: Adjustment to the city-identity-sustainability network structural model, *Environment and Behavior*, **34**(1), 67-80.

Proshansky, H.M., Fabian, A.K. & Kaminoff, R. (1983) Place-identity: Physical world socialization of the self, *Journal of Environmental Psychology*, **3**, 57–83.

Relph, E. (1976) *Place and Placelessness*, London: Pion.

Richards, G. (2017) From place branding to placemaking: The role of events, *International Journal of Event and Festival Management*, **8**(1), 8-23.

Schnell, S.M. (2013) Deliberate identities: Becoming local in America in a global age, *Journal of Cultural Geography*, **30**(1), 55-89.

Shortridge, J. (1996) Keeping tabs on Kansas: Reflections on regionally based field study, *Journal of Cultural Geography*, **16**(1), 5-16.

Sims, R. (2009) Food, place and authenticity: Local food and the sustainable tourism experience, *Journal of Sustainable Tourism*, **17**(3), 321-336.

Stokowski, P.A. (1991) Sense of place as a social construct, paper presented at the 1991 National Recreation and Parks Association Research Symposium, 17-20 October, Baltimore, MD.

Stone, L. (1989) Cultural cross-roads of community participation in development: A case from Nepal, *Human Organisation*, **48**, 206-213.

Strzelecka, M. (2012) Individual and Community Empowerment Enhancement in Sustainable Tourism Development in Post-Communist Poland, PhD thesis, University of Illinois at Urbana-Champaign, Illinois.

Strzelecka, M., Sorensen, J. & Wicks, B. (2011) The role of place-attachment in revitalization of neighborhood parks in East St. Louis, *Society and Leisure*, **33**, 251-271.

Theodori, G. (2005) Community and community development in resource-based areas: Operational definitions rooted in an interactional perspective, *Society and Natural Resources*, **18**, 661–669.

Timothy, D.J. & Ron, A.S. (2013) Heritage cuisines, regional identity and sustainable tourism, in C.M. Hall & S. Gossling (eds.), *Sustainable Culinary Systems: Local Foods, Innovation, and Tourism and Hospitality*, London: Routledge, pp. 275-290.

Tuan, Y-F. (1974) Space and place: Humanistic perspective. *Progress in Human Geography*, **6**, 233-246.

Tuan, Y-F. (1977) *Space and Place: The Perspective of Experience*, Minneapolis, MN: University of Minnesota Press.

Vallance, S., Perkins, H.C. & Dixon, J.E. (2011) What is social sustainability? A clarification of concepts, *Geoforum*, **42**, 342-348.

Vaske, J.J. & Kobrin, K.C. (2001) Place-attachment and environmentally responsible behavior, *Journal of Environmental Education*, **32**(4), 16-21.

Wilkinson, K.P. (1972) A field theory perspective for community development research, *Rural Sociology*, **37**(1), 43-52.

Wilkinson, K.P. (1991) *The Community in Rural America*, Middleton, WI: Social Ecology Press.

Williams, D. & Vaske, J. (2003) The measurement of place-attachment: Validity and generalizability of a psychometric approach, *Forest Science*, **49**(6), 830–840.

Williams, D.R., Patterson, M.E., Roggenbuck, J.W. & Watson, A.E. (1992) Beyond the commodity metaphor: Examining emotional and symbolic attachment to place, *Leisure Sciences*, **14**, 29–46.

10 Tourism in the Intermountain West: The Role of Food and Agritourism in Neolocalism

Kynda R. Curtis, Miranda Bradshaw and Susan L. Slocum

Abstract

The strong tourism industry in the Intermountain West – coupled with its growing local foods movement – provides fertile ground for food tourism expansion. This study examines non-resident tourist characteristics, attitudes, and interests related to food tourism activities through a visitor survey. Specifically, we examine tourists' associations among a variety of food and agritourism activities. We then group tourists into target markets to better understand their characteristics and interests. We discuss the primary target markets including preferred activities and best practices for promotion. Additionally, we provide suggestions for destination design based on three distinct groupings of food tourism related activities.

Introduction

Tourism is among the most important and largest industries in the western USA, supported by a network of national parks, outdoor recreation areas, state parks, ski resorts, festivals and special events, premier wine growing regions, and many historic and religious sites. Visitor spending drives economic activity in the parks, recreation, and hospitality sectors across the West, and tourism is responsible for job creation and increasing levels of tax revenue (Utah Economic Council, 2015). The Intermountain West of the USA – the geographical region between the Rocky Mountains to the east and the Cascade and Sierra Nevada ranges to the west – boasts some of the nation's leading national parks and winter recreation activities. Annual ski visits to the region in 2017 numbered 20.8 million (RRC Associates, 2019), while US National Park Service visitations numbered 57.5 million (Ziesler & Singh, 2017). Table 10.1 summarises tourism activity for four states in the Intermountain West.

Table 10.1: Intermountain West tourism statistics – selected states (2017)

	Utah	Idaho	Nevada	Colorado
Total visitors (millions)	19.3	10.9	31.6	84.7
Visitor spending (billions $USD)	9.15	4	63	20.9
Jobs supported	147,000	35,000	458,780	171,000
Ski visits (millions)	4.15	Unknown	Unknown	7.1
National park visits (millions)	10.5	Unknown	2.9	7.5
National Park Service (NPS) visits (millions)	15.1	0.7	6	7.6

Sources: Colorado.com (2018), Idaho Commerce (2018), Ziesler and Singh (2017), RRC Associates (2019), Nevada Industry Partners (2018), and Utah Office of Tourism (2018).

In Utah, tourism is a well-developed, $9.15 billion industry (Utah Office of Tourism, 2018). In 2017, over 19 million people visited Utah, primarily to go skiing (4.15 million visitors) and to visit national parks (10.5 million visitors). Utah is home to the 'mighty five' national parks—including Zion, Bryce Canyon, Capital Reef, Arches, and Canyon Lands. Utah's primary port of entry, the Salt Lake City International Airport, served 25.5 million passengers in 2018, making it the 23rd busiest airport in the USA (SLC International Airport, 2019). While tourism is Utah's second largest industry overall, tourism plays an even more vital role in Utah's rural communities, where tourism-related jobs can account for up to 40% of the job market (Leaver, 2014).

Growing consumer interest in food origins and production processes has given rise to the 'buy local' or 'locavore' movement, which emphasises purchasing food grown and/or processed close to home, such as in-state or within a certain number of miles (Slocum & Curtis, 2018). Evidence of the growing locavore movement in the USA is reflected in direct-to-consumer sales, which have exhibited positive growth over the past three decades. For example, between 1978 and 2012, the total value of direct-to-consumer sales increased by nearly 240% (Low et al., 2015). Additionally, there was a 200% increase in farmers' markets from 2000 to 2016, a 275% increase in consumer supported agriculture programs from 2004 to 2014, and a 288% increase in food hubs from 2007 to 2014.

The locavore movement plays an important role in the growing food tourism industry and can support the neolocal process of crafting the tourism product around traditional agricultural practices. Specifically: dining out; visiting farmers' markets, breweries and wineries; and participating in agritourism activities while traveling can enhance sense of place and provide enticing narratives that draw tourists to rural agricultural communities. The National Restaurant Association (2015) reported that seven out of ten consumers were more likely to visit a restaurant offering locally sourced items. Food and culinary tourism are increasingly popular, as 27 million travellers participated in wine or culinary-related activities over a 3-year period (Sohn & Yuan, 2013).

At the same time, agricultural producers, especially small producers in rural areas, face unique challenges and may benefit from food and agritourism development and promotional efforts directed toward food tourists (Bond et al., 2008). Mass agricultural production is often governed by strict contracts and middlemen, leading to disproportionately small revenues for producers. Unreliable weather patterns, capricious growing seasons and fluctuating markets combine for uncertain yields, prices, and profits (Fleisher, 1990). Direct marketing agricultural goods through roadside stands, farm shops, farmers' markets, and restaurants to residents and tourists, when marketed appropriately, can improve profits for small producers with entrepreneurial skills (Slocum & Curtis, 2017). Destination marketing messages around food and agritourism products encourage tourists to spend their budgets in a way that reduces economic leakages (Getz et al., 2014).

The high quality and diversity of Utah's agriculture and food, combined with Utah's popularity as a vacation destination, may provide unique opportunities for the neolocal development of food and culinary tourism. The effective marketing of local foods to tourists may be a powerful rural economic development strategy, connecting Utah's flourishing tourism industry with local food providers and agritourism operators. Agritourism, especially, has been shown to have cultural, economic, and environmental benefits (Barbieri et al., 2019). However, understanding food tourists' characteristics, motives, and behaviours is key to attracting them (Getz et al., 2014). Thus, this study analyses Utah tourists using a 2014 in-person survey of 700 non-resident visitors examining the types of experiences and activities they seek, how they research and plan their travel, the reason for their visit, their spending habits, length of trip, attitudes toward Utah as a destination, and propensity to make a return visit and recommend Utah to others. We use factor analysis to identify key food- and agritourism-related activities of interest to Utah tourists. We then use cluster analysis to group tourists into target markets according to demographics, interests, attitudes, and trip characteristics. The study results may be used by both the agriculture and tourism industries to support neolocal place-based products, develop targeted marketing plans, and design destinations and tourism products to best suit tourists seeking authentic food and agritourism experiences in the region.

Literature review

Neolocalism is the process of asserting a distinctive brand image based on local attributes and sense of place in an attempt to communicate culture and heritage (Flack, 1997). As a relatively new approach to marketing tourism products, there is very little research on consumer markets and customer attributes that support neolocalism. Research suggests that neolocal consumers are interested in cultural tourism products (Richards, 2011) and may align within sustainable consumption patterns. It is understood that consumers of neolocalism demand a deeper connection with destinations, often in response to the marketing of specific products that promise authentic and unique experiences (Rudy, 2004). Recognition of neolocalism in tourism occurred through the emerging craft beer scene, which implies that a better understanding of food tourists has the potential to inform neolocal consumption practices (Goodman et al., 2011). As Rogerson (2016: 17) writes, 'the rise of alternative food networks is the most visible expression of neolocalism'.

Defining food tourism or profiling so-called food tourists is challenging due to a lack consensus around the definition of food tourism and the determinants of 'food tourists' (Martinez et al., 2010). Hall et al. (2011: 10) describe food tourism as the 'visitation to primary and secondary food producers, food festivals, restaurants, and specific locations for which food tasting and/or experiencing the attributes of a specialist food production region are the primary motivating factor for travel'.

Long (2004: 21) defines culinary tourism as 'the intentional, exploratory participation in the foodways of another – including the consumption, preparation, and presentation of a food item, cuisine, meal system, or eating style considered to belong to a culinary system not one's own'.

Food tourism takes the shape of a variety of activities and experiences including (but not limited to) guided tours of farms or wineries, foodways, cooking classes, food and wine festivals, eating at local-sourcing restaurants, and shopping at farmers' markets or other direct-to-consumer outlets (Smith et al., 2010). Improving local food offerings in a region may lead to an improved destination image (Getz et al., 2014). Yun et al. (2011: 11) note, 'product development of culinary tourism must be enjoyable, easy to consume, and presented in a manner that is connected to other cultural activities and attractions in order to achieve its maximum positive impact on tourists, producers and the community'. Not only does food tourism attract those interested in food-related experiences, it also encourages those visiting for other reasons to spend their food budgets in ways that may assist local producers (Getz et al., 2014).

While many tourists' primary travel motivation is not to participate in food tourism, all tourists must eat and many become incidental food tourists (Hall et al., 2011; Yun et al., 2011). While many people appreciate quality food, Getz et al. (2014) explain that 'foodies' are self-declared food experience seekers who choose to consume local, sustainably grown foods and whose lifestyle and mindset toward travel differ from the typical tourist. Although participating in the locavore movement does not necessarily make one a 'foodie' or 'food tourist', the recent emergence of the foodie movement may indicate that locality is a salient food attribute for some consumer segments. Buratti (2019) believes that consumers of neolocalism want products embedded in place that represent local values and employ local resources, such as local food networks. Similarly, the appeal, availability and authenticity of local foods are important aspects of attracting neolocal tourists through agritourism ventures (Curtis & Monson, 2004; Getz et al., 2014).

Despite the lack of a widely accepted definition of food tourism, profiling food tourists may lead to a better understanding of exactly how the needs, expectations, and behaviours of food-minded tourists differ from the average tourist (Wolf, 2014). Food tourists are typically more experienced travellers between the ages of 30 and 60 and are equally represented among men and women (Wolf, 2014). Food tourists tend to be more educated and earn higher incomes than the average non-food-oriented tourist (Shenoy, 2005). Shenoy (2005) established five factors underlying participation in food tourism, which were labelled as 'purchase local', 'dine local', 'drink local', 'dine elite', and 'familiarity'. These factors were then used in a cluster analysis to group respondents. The groups were labelled as culinary tourists, experiential tourists, or general tourists, based on differences in behaviours related to food tourism. The culinary tourist cluster were respondents who ranked high on the dine local and drink local factors. The experiential tourist cluster ranked high on dine local but also ranked high on the familiarity factor. General tourists scored fairly evenly across the five factors.

Sánchez-Cañizares and López-Guzmán (2012) profiled tourists into consumer segments according to the role that food and gastronomy played in their visit to Córdoba, Spain. Results indicated that visitors whose primary reason for visiting the city was gastronomy were more likely to be males who were familiar with local wines. They also found significant differences in tourists' enjoyment of attractions in Córdoba, their satisfaction with gastronomy, and overall satisfaction with their trip according to which consumer segment the respondent belonged.

Agritourism includes a variety of activities including farm stays, 'pick your own'/u-pick activities, farm tours, farmers' markets, hay rides, pumpkin patches, corn mazes, and recreational activities such as fishing, hunting, and horseback riding (Lobo et al., 1999; Slocum & Curtis, 2018). Wilson et al. (2006: 1) define agritourism as 'anything that connects consumers with the heritage, natural resources, or culinary experiences unique to the agricultural industry, or a particular region of the country's rural areas'. Agritourism incorporates visitation to working farms, as well as commercial on-farm tourism services.

A key motive for participating in agritourism may be a desire to explore the natural environment rather than commercially developed tourist attractions (Curtis & Monson, 2004), which is well aligned with the motivations of neolocal consumers (Murray & O'Neill, 2012). A survey conducted by the US Department of Agriculture Forest Service (2002) found that 71% of respondents wanted to learn where food comes from, 64% wanted to participate in farm activities, 43% wanted to pick fruit or produce, 39% wanted to purchase agricultural products, and 27% wanted to hunt or fish. Understanding the role that agritourism plays in trip motivation and the activities sought after by tourists is important to understanding how to promote and develop agritourism opportunities offered by farmers, ranchers, and communities (Gascoigne et al., 2008).

Researchers have found that a variety of sociodemographic factors correlate with the demand for agritourism activities. Carpio et al. (2008) found that race and location of residency had the greatest impact in determining the number of farm trips taken per visitor per year. They found that white respondents were 10% more likely to visit a farm, relative to the baseline respondent, whereas Hispanic respondents were 13% less likely to visit a farm. They also found that someone living in an urban area is 5% less likely to visit a farm than the baseline respondent.

Age, marital status and family characteristics may also be important. For instance, in a survey of Colorado agritourists, the average agritourist was 46 years old, 73% were married, and 28% were young couples without children (Thilmany et al., 2007). Qualitative comparisons were also made between in-state and out-of-state travellers, travel frequency, and participation in agritourism activities. Barbieri and Tew (2010) surveyed farms to discover the common characteristics of agritourism visitors in Missouri. They found that agritourists tend to be senior citizens or families with young children. These findings suggest that consumers interested in agritourism may be a highly heterogeneous group.

Survey data overview

Data for this study were collected in 2014 through an in-person survey, administered by trained surveyors across Utah. Data collection sites included ski areas, airports, national parks, and convention and visitor information centres. Every third person was approached and asked whether they would like to participate in the survey. A total of 700 fully completed surveys were used in the analysis. Only those responses collected from tourists whose permanent residence was outside of Utah were included.

Survey participants were asked to answer a variety of questions designed to assess their participation in agricultural and food activities and events, such as participation in community supported agriculture, farmers' markets, and food or wine festival attendance, both when at home and while traveling. Other questions targeted the nature of the participants' experience in Utah, including the kind of establishment they stayed in, the number of times they had previously visited, and the duration of their stay. There were questions focused on the participants' motivation for visiting Utah and their primary leisure interests. The participants' destination image of Utah culture, food, and tourist activities were also assessed through a variety of questions. Question types included basic demographics, short answer, dichotomous choice, and multiple choice. Table 10.2 provides sample summary statistics.

Table 10.2: Sample descriptive statistics

Item	%age of Sample	Mean	N
Demographics			
Age (years)		50 (17.05)	635
Gender			668
Male	52		
Female	48		
Marital status			665
Single	32		
Married	68		
Ethnic background			673
Caucasian	84		
Asian	5		
Hispanic	4		
Other	2		
Income (annual, $)		103,151 (131,348)	376
Education			671
No college	29		
Bachelor's degree	31		
Graduate degree	40		
Employment			672
Full-time employed	49		
Part-time employed	10		
Unemployed	4		
Homemaker	4		
Student	4		
Retired	29		
Trip characteristics			
Length of stay		10.63 (21.6)	689
Number of previous visits			692
None	29		
1–3 times	36		
4–6 times	10		
7–11 times	7		
12 or more times	18		

Item	%age of Sample	Mean	N
Primary reason for trip			611
Business	1		
Family/friends	5		
National parks	9		
Outdoor recreation	43		
Cultural/heritage sites	24		
Special event/festival	2		
Agritourism	9		
Passing through	6		
Accommodations			700
Hotel/motel	54		
Bed and breakfast	2		
Camping	20		
Resort	2		
Family and friends	10		
Other	13		
Travel party characteristics			
Party size			689
Adults		2.9 (3.41)	
Children		1.56 (1.49)	
Resources used in trip planning			675
Internet/website	41		
Brochure/booklet	10		
Recommendation	3		
Tradition	32		
Other	14		
Spending habits			
Expense per person ($)		571.70 (835.50)	689
Percentage of budget spent on food (%)		18 (17)	645

Survey sample statistics show that 52% of respondents were men and 68% were married. The average respondent was 50 years old. About 60% of participants were between the ages of 40 and 70. The vast majority of respondents were Caucasian (84%), with 5% and 4% identifying as Asian and Hispanic, respectively. Most respondents had a college degree (71%), and 49% were employed full-time. The average annual household income in 2013 was about $103,000.

The most common primary reasons for traveling to Utah were outdoor recreation (43%), visiting cultural and heritage sites (24%), national parks (9%), and agritourism (9%). Only 1% of respondents indicated that business was the primary reason for travel. On average, visitors stayed 10.6 days and were most likely to stay in hotels or motels (54%), although 20% indicated that they were camping. Of the visitors, 29% had never been to Utah before, while 35% had visited Utah at least four times previously. In planning their trip, the Internet was the most popular resource (41%), with only 10% using brochures or booklets. Nearly one third (32%) of respondents stated that their trip was based on tradition. Visitors travelled in small groups, typically consisting of about three adults (2.9) and one to two children (1.5). Nearly one-third of the travel groups (30%) included children under the age of 18. On average, travellers spent about $570 per person for the duration of their trip. Of the total trip budget, about 18% was spent on food.

Respondents were asked to rank, on a scale of 1 to 5, their level of participation in food-related activities when at home and while traveling. Low rankings corresponded with less participation and high rankings corresponded with more participation. Tables 10.3 and 10.4 summarise the responses. When at home, respondents showed strong preferences for buying foods from local producers, cooking and trying new foods, and recycling. Among alternative agriculture activities, buying locally grown foods (3.52) and shopping at farmers' markets (3.08) ranked the highest among respondents. Among food-related activities, cooking at home (4.29), trying new food items or recipes (3.86), and eating ethnic foods (3.57) were all ranked highly. Food canning (1.80) and beer and wine making (1.35) were the least popular food-related activities. Recycling (4.35) was the highest ranked sustainable activity and the highest ranked activity overall.

While traveling, respondents also showed noticeable preferences for certain food-related activities. Trying new food items (3.48), trying local recipes (3.12), and seeking out restaurants that source locally (2.97) were the highest ranked food-related activities. Buying locally sourced food (2.80) and shopping at farmers' markets (2.46) were the most popular agriculture-related activities. Once again, recycling (3.67) was the highest ranked activity overall.

Table 10.3: Sample at-home activities

When at home, how often do you participate in the following activities? (1 = Never, 2 =Rarely, 3 = Sometimes, 4 = Often, 5 = Always)		
Activity	**Mean(SD)**	**N**
Alternative agriculture		
Buy locally sourced food	3.52 (1.16)	666
Shop at farmers' markets	3.08 (1.14)	667
Participate in consumer supported agriculture	1.73 (1.11)	627
Buy organic certified produce	2.79 (1.19)	659
Visit local farms	2.05 (1.14)	652
Alternative foods		
Cook at home	4.29 (0.80)	665
Try new food items or recipes	3.86 (0.91)	659
Buy unfamiliar produce	2.93 (1.19)	659
Eat ethnic foods	3.57 (1.05)	655
Attend beer/wine festivals	2.39 (1.27)	658
Food canning	1.80 (1.11)	652
Beer/wine making	1.35 (0.88)	648
Sustainable activities		
Home gardening	2.62 (1.50)	656
Composting	2.14 (1.49)	655
Recycling	4.35 (1.07)	654

Table 10.4: Sample while-traveling activities

While traveling, how often do you participate in the following activities? (1 = Never, 2 =Rarely, 3 = Sometimes, 4 = Often, 5 = Always)		
Activity	**Mean(SD)**	**N**
Alternative agriculture		
Buy locally sourced food	2.80 (1.19)	620
Shop at farmers' markets	2.46 (1.16)	624
Visit local farms	1.77 (0.95)	607
Spend the night at local farms	1.28 (0.65)	609
Participate in agritourism	1.60 (0.88)	606
Alternative foods		
Cook at accommodations	2.82 (1.31)	635
Try new food items	3.48 (1.06)	317
Try local recipes	3.12 (1.19)	610
Buy food items as souvenirs or gifts	2.72 (1.13)	622
Seek out restaurants that source locally	2.97 (1.23)	623
Attend beer/wine festivals	2.22 (1.28)	620
Sustainable activities		
Recycling	3.67 (1.33)	622

On a similarly designed scale, respondents were asked to rank their likelihood of returning to Utah, their agreement with several characterisations of Utah's destination image, and certain interests and experiences regarding their trip to Utah (see Table 10.5). Respondents felt that Utah is best known for its outdoor activities (4.48) and its landscapes (4.63). While respondents indicated that they had seen food advertised as 'Utah's Own' or 'Local First Utah' (3.89), they felt less strongly about how well local food was advertised (2.94) and were less likely to feel that the food they had eaten in Utah was good (2.73). Among statements related to satisfaction, respondents indicated that they planned to return to Utah (4.10) and would recommend Utah to their family and friends (4.43). Respondents generally disagreed with the statement that they had sufficient time to see and do everything they wanted while in Utah.

Table 10.5: Sample opinions and interests

Please state your level of agreement with the following statements (1 = Strongly disagree, 2 = Disagree, 3 = Unsure, 4 = Agree, 5 = Strongly Agree)		
Item	**Mean(SD)**	**N**
Destination image		
Utah is known for its outdoor activities (skiing, hiking)	4.48 (0.69)	678
Utah is known for its landscapes (national parks, wilderness)	4.63 (0.61)	679
Utah is known for its heritage and culture	3.86 (0.89)	666
Utah has a strong food culture	3.65 (1.00)	664
Local products and attractions		
The food I have eaten in Utah is good	2.73 (1.27)	665
Local food is well advertised in Utah	2.94 (0.79)	665
I have seen food advertised as 'Utah's Own' or 'Local First Utah'	3.89 (0.74)	662
Community-based attractions are well advertised in Utah	2.94 (0.87)	658
Locally produced crafts were readily available in tourist shops	2.50 (1.07)	655
Experiences, interests, and satisfaction		
I am interested in learning more about Native American culture	3.35 (0.93)	653
I am interested in learning more about Mormon culture	3.36 (0.89)	650
I feel I have an understanding of Utah culture	3.16 (0.96)	657
I had enough time to see and do everything I wanted while in Utah	2.79 (1.21)	658
I am planning to return to Utah	4.10 (0.89)	672
I will recommend Utah as a travel destination to my family and friends	4.43 (0.69)	672

Methods

Factor analysis

We employed factor analysis to identify food and agritourism activities closely related in the minds of Utah tourists. Factor analysis is a statistical technique used to find underlying dimensions or 'factors' that describe the primary sources of variation between variables. Factor analysis simplifies data interpretation by reducing a large number of variables to a smaller number of factors that represent the multidimensional decisions often made by consumers. Each individual factor has a corresponding factor weight that can be described as the relationship between the variable and its corresponding factor (Bond et al., 2008).

We performed factor analysis on 27 items representing activities associated with food and agritourism, both while traveling and when at home (see Tables 10.3 and 10.4). The Kaiser–Mayer–Olkin test (Cerny & Kaiser, 1977) of sampling adequacy (0.8) indicated that the various activities shared enough variance to warrant the application of factor analysis. Principal component factor analysis using varimax rotation and the Kaiser (1958) criterion was used. Only factors with an associated eigenvalue greater than or equal to 1 were retained.

In the factor analysis for the 'at-home' variables, four factors emerged, which cumulatively explained 53.59% of the total variance and were named 'local foods', 'food experiences', 'do it yourself', and 'food connections'. The factor analysis for the 'while-traveling' variables returned three factors that together explained 54.48% of the total variance; these were named 'food tourism', 'agritourism', and 'local foods'. Tables 10.6 and 10.7 list the factors and their associated activities (variables).

Each factor contains several related activities, or dimensions, which implies that if a traveller participates in one activity of the factor, they are also likely to participate in the other activities in the same factor. Understanding how tourists view various activities as part of an overall experience is essential for successful tourism destination design.

Cluster analysis

We used a partition clustering method to explore the natural target markets occurring based on the similarity of participant responses to a variety of survey questions. The cluster analysis draws upon responses to the questions included in Table 10.2, as well as financial information provided by respondents. Before clustering the data, the variables

Table 10.6: Factor analysis results for at-home activities (N = 571)

Activity	Factor Loading	Eigenvalue	Percentage Variance Explained
Local foods		3.917	8.58
Shop at farmers' markets	0.638		
Buy organic produce	0.61		
Visit farms	0.59		
Food experiences		1.575	10.5
Try new foods/recipes	0.761		
Eat ethnic foods	0.728		
Try new produce	0.681		
Do it yourself		1.287	26.11
Gardening	0.795		
Canning/preserving	0.747		
Composting	0.667		
Food connections		1.259	8.39
Recycle	0.721		
Cook at home	0.541		
Buy local foods	0.369		
Total variance explained			53.59

Table 10.7: Factor analysis results for while-traveling activities (N = 571)

Activity	Factor Loading	Eigenvalue	Percentage Variance Explained
Food tourism		3.923	32.69
Try new foods	0.784		
Try local recipes	0.751		
Buy food-related gifts	0.571		
Agritourism		1.484	12.37
Spend a night at a farm	0.734		
Agritourism activities	0.722		
Visit farms	0.337		
Local foods		1.13	0.42
Buy local foods	0.639		
Cook at accommodations	0.632		
Shop at farmers' markets	0.377		
Total variance explained			54.48

were standardised (StataCorp, 2011). Clustering was performed using k-means, which performs a partition cluster analysis. A random-number seed was used to randomly select observations from across the range of the data to form initial group centres. Observations were then assigned to mutually exclusive groups based on the amount of similarity between the mean of the observation and the group mean (StataCorp, 2011).

We used the Calinski–Harabasz (1974) pseudo-F index to determine the optimal number of groups. The highest Calinski–Harabasz pseudo-F score (19.5) occurred when the cluster analysis formed six clusters. However, two of the clusters contained only 5% of the total number of observations each, which seemed unlikely to result in truly beneficial information for tourism industry providers, agricultural producers, or policy-makers. For the sake of useful interpretation of the data, we instead selected a cluster analysis forming four clusters (see Table 10.8). Each cluster represents a group of Utah visitors with characteristics distinct from the other clusters.

Table 10.8: Cluster analysis results

Cluster	Name	N	Percentage
1	Large family trips	36	8.6
2	Outdoor enthusiasts	123	29.3
3	Sophisticated food travellers	125	29.8
4	Food and culture pilgrims	91	21.67
Total		375	89.37

A brief summary of the most distinct characteristics of each group are highlighted. Cluster 1, large family trip (9%): The average age of this group was 52 years; 67% were married and well educated, as 14% had a graduate degree. They travelled in the comparatively largest groups (4.8 people on average) and had visited Utah many times, staying 5 days on average. About 52% used the Internet in trip planning and were unlikely to use brochures. This group was the most likely to participate in agri-tourism activities, rather than outdoor recreation or visiting national parks, while traveling and to participate in canning, composting, and gardening at home. The characteristics of this group and their interest in agritourism are opposite to those found in two studies conducted in Colorado (Gascoigne et al., 2008, Thilmany et al., 2007), which found small groups without children were more likely to focus on agritourism. However, results are consistent with Barbieri and Tew (2010), who found that families with young children are likely to participate in agritourism.

Cluster 2, outdoor enthusiasts (29%): This group tended to be more male than female, was the least likely to be married and travelled in the smallest groups (average of 2.8 people). Comparatively, they were the least educated and had an average age of 48 years. They spent the least amount of time in Utah (average of 4.7 days) and were the most likely to have used a brochure in trip planning. They were highly interested in outdoor recreation. They spent about 22% of the total trip budget on food but had the smallest overall trip budget of all the groups. This group was the least likely to seek out local foods while traveling. While Sánchez-Cañizares and López-Guzmán (2012) found that younger males sought culinary experiences in Spain, the differences in culture and study focus likely explain the contradictory outcomes.

Cluster 3, sophisticated food travellers (30%): This was the oldest group (average age was 53) and the most likely to be married. Nearly 60% had graduate degrees, making this the most educated group. Their average trip lasted 14.3 days, and 74% used the Internet in trip planning. This group was the most likely to travel for business reasons, but they also visited national parks and cultural and heritage sites. They spent about 22% of the total trip budget on food. This group was the most likely to seek out local foods while traveling and participated in sustainable activities (recycling, etc.) at home. At home, they do a lot of their own cooking and buy local foods. The results for this group are consistent with the profile of the typical food tourist as shown in Shenoy (2005) and Wolf (2014).

Cluster 4, food and culture pilgrims (22%): This group was the youngest, and about 64% were married and female. This group was the most likely to have a bachelor's degree and travelled in groups of about 3.8 people. The least likely to use the Internet in trip planning, this group travelled mostly based on tradition and were likely to have visited Utah previously. This group tended to visit cultural and heritage sites. They spent about 15% of the trip budget on food and spent the most overall per person ($960) on a typical 10.5-day vacation. This group was the most likely to seek out food experiences, buy food souvenirs or gifts while traveling, and to try new foods and local foods when at home. They often belonged to beer and wine clubs. The profile of this group – younger adults with one or two children – exemplifies a typical return travel situation: repeated visits to preferred locations, eating at favourite restaurants, eating regional foods and bringing back coveted local goods to family and friends at home. These emotional connections to a destination are a primary motivator for return visitation (Oom Do Valle et al., 2008).

Discussion

Factor analysis was conducted on distinct food and agritourism activities respondents participate in at home and/or while traveling, which resulted in four distinct factors for at-home activities and three for while-traveling activities. Each factor contains a set of activities that Utah visitors associate together or view as similar. Thus, if they participate in one of the activities, they are likely to want to participate in another activity in the same factor. Hence, tourism destinations that provide several activities will bestow higher tourist satisfaction than an experience that misses one or more of those interests.

The food tourism factor appears to best support neolocal food-related activities. Their interest in culinary experiences, such as ethnic food cooking classes, provide an opportunity to explore new foods, local recipes, and opportunity to purchase food souvenirs or gifts (jam, cheese, honey, etc.). This group would also support a farm shop or roadside stand where destinations can highlight regional foods, provide local recipes using those foods, and offer pre-made meals typical of regional cuisine or other packaged foods for tourists to sample or purchase as souvenirs.

Neolocal tourists looking to visit farms would also prefer to lodge there and participate in activities on the farm. Thus, the agritourism factor would indicate that agritourism operations might consider offering lodging, such as cottages, camping space, etc. to provide a more holistic experience for the tourist. Another example might be a raspberry u-pick operation that adds a café or tearoom offering raspberry shakes, smoothies, or fresh baked goods to expand its offerings and the variety of activities provided. The local foods factor, which includes purchasing local foods, shopping at farmers' markets and cooking at accommodations, shows that farmers' market vendors who prominently display local offerings, provide recipes, or sell complete ingredient packages for recipes, so that cooking at visitor accommodations is easier, will enhance the visitor experience.

The cluster analysis was used to group respondents with similar characteristics (demographics, interests, perceptions, activities, and travel motivations), such that each group represents a distinct target market.

The four primary groups are large family trip, outdoor enthusiasts, sophisticated food travellers, and food and culture pilgrims.

The characteristics of the large family trip group suggest that agritourism activities developed in a manner that is educational and friendly to families or other large groups may target this group particularly well. Agricultural producers considering the development of agritourism opportunities as a part of their operations should consider activities that are conducive to large groups and a wide range of ages, such as hayrides, educational demonstrations, and farm tours. In instances where one activity is more conducive to a particular age group, complementary activities may be paired to appeal to a wider range of ages. An example could be pairing a corn maze, which may be less appealing to some older or younger tourists, with a hayride or educational demonstration. Online advertisement is likely an effective way to market activities to this group.

For outdoor enthusiasts, outdoor recreation is their primary travel motivation, and they are the least interested in food tourism of the four groups. However, local producers may still benefit by promoting food-related experiences and food products to this group. Expos, conventions, and festivals related to outdoor recreation may be a good venue for exploring traveller interest in local foods tailored to their specific needs. For example, producers may find that this group is interested in foods that can be prepared in the outdoors or that can be transported and stored easily, such as jerky, dried fruit, or other dehydrated foods.

The sophisticated food travellers group was the most likely to travel for business reasons but also travelled to visit national parks and cultural and heritage sites. Because these travellers were the most likely to use the Internet in trip planning, tourism providers and local producers should consider using the Internet to develop and promote tours, restaurants, farm experiences, and local foods targeted at this group. Advertising and developing culinary tourism opportunities in connection with other tourist attractions may be an effective way to interact with this group. For example, a tourist experience combining a food, wine, or beer tasting with a museum tour, natural history tour, or historic site may suit the needs and interests of this group. Since this is the oldest tourist segment, comparatively, limited mobility and other age-related factors may need to be considered when developing tourism experiences. The sophisticated food travellers group also participates in sustainable activities when at home. Therefore, presenting culinary tourism opportunities as a part of sustainable tourism or a sustainable lifestyle may be another way to attract members of this group. For example, a local-sourcing restaurant may appeal to this segment by seeking to develop a pro-sustainability reputation. This may be accomplished in a variety of ways, including local and online advertising, participation in educational community events, or developing 'meet the grower' type activities.

The food and culture pilgrims group was the youngest group and primarily married and female. This group is made up of repeat tourists whose travel planning is mostly based on tradition. They are the most likely to visit cultural and heritage sites, spend the most per person relative to the other groups, and are the most likely to belong to beer and wine clubs. The development and promotion of cultural and heritage-related activities alongside wine tastings, beer tastings, and drink-related festivals, may be an effective way to interact with this group. For example, a tour of a historic building in downtown Salt Lake City, Utah, that culminates in a tasting of locally produced foods and beers would represent an activity specifically tailored to this group's interests. This group is also likely to buy food-related gifts. Therefore, the availability and transportability of food-related gifts at the site of cultural and heritage-related attractions and activities may provide an opportunity for local producers to market their products to this tourist segment.

Conclusion

The strong tourism industry in the Intermountain West – coupled with its growing local foods movement and neolocal marketing focus – provides fertile ground for food and agritourism expansion in the region. As understanding tourists' characteristics, motives, and behaviours is key to attracting them, thus, this study examined non-resident tourist characteristics, attitudes, and interests related to food and agritourism activities.

We conducted two types of analysis. First, factor analysis was used to examine how tourists associate a variety of food and agritourism activities so that activities with close associations can be combined to provide a complete destination offering. The results provide three distinct clusters, or groupings, of food- and agritourism-related activities. Destination offering suggestions were provided for each based upon results.

Second, cluster analysis was performed to group tourists into target markets to better understand their characteristics and interests in order to improve destination design and effectively promote offerings to each market. We found four primary target markets among current tourists. While two of the four markets activity sought out food and culinary activities and a third, agritourism activities while traveling, all four markets provided evidence of neolocalism and were good candidates for food and culinary experiences. The characteristics of each market and examples on how best to promote and cater to them were discussed.

References

Barbieri, C. & Tew, C. (2010) The economic benefits of agritourism in Missouri farms, report prepared for the Missouri Department of Agriculture.

Barbieri, C., Sotomayor, S. & Aguilar, F.X. (2019) Perceived benefits of agricultural lands offering agritourism, *Tourism Planning and Development*, **16**(1), 43-60.

Bond, C.A., Thilmany, D. & Bond, J.K. (2008) Understanding consumer interest in product and process-based attributes for fresh produce, *Agribusiness: An International Journal*, **24**(2), 231-252.

Buratti, J.P. (2019) A geographic framework for assessing neolocalism: The Case of Texas Hard Cider, PhD dissertation, Texas State University, San Marcos.

Calinski, T. & Harabasz, J. (1974) A dendrite method for cluster analysis, *Communications in Statistics – Theory and Methods*, **3**, 1-27.

Carpio, C.E., Wohlgenant, M.K. & Boonsaeng, T. (2008) The demand for agritourism in the United States, *Journal of Agricultural and Resource Economics*, **33**(2), 254-269.

Cerny, C.A. & Kaiser, H.F. (1977) A study of a measure of sampling adequacy for factor-analytic correlation matrices, *Multivariate Behavioral Research*, **12**(1), 43-47.

Colorado.com (CCL). (2018) Colorado tourism sets all-time records for eighth consecutive year. Available at https://www.colorado.com/news/colorado-tourism-sets-all-time-records-eighth-consecutive-year. (Accessed: 10/09/2018).

Curtis, K.R. & Monson, J. (2004) *Agritourism: Opportunity for farm diversification in Nevada*, Extension fact sheet prepared for the University of Nevada Reno Extension.

Flack, W. (1997) American microbreweries and neolocalism: "Ale-ing" for a sense of place, *Journal of Cultural Geography*, **16**(2), 37-53.

Fleisher, B. (1990) *Agricultural Risk Management*, Boulder, CO: Lynne Rienner Publishers.

Gascoigne, W., Sullins, M. & McFadden, D.T. (2008) Agritourism in the West: Exploring the behavior of Colorado farm and ranch visitors, *Western Economics Forum*, **7**(2), 12-24.

Getz, D., Andersson, T. & Vujicic, S. (2014) *Foodies & Food Tourism*, Oxford: Goodfellow.

Goodman, D., DuPuis, E.M. & Goodman, M.K. (2011) *Alternative Food Networks: Knowledge, Practice and Politics*, New York, NY: Routledge.

Hall, C.M., Sharples, L., Mitchell, R., Macionis, N. & Cambourne, B. (2011) *Food Tourism around the World: Development, Management, and Markets*, Abingdon: Routledge.

Idaho Commerce. (2018) Tourism resources. Available at https://commerce.idaho.gov/tourism-resources/. (Accessed: 15/07/2019).

Kaiser, H.F. (1958) The varimax criterion for analytic rotation in factor analysis, *Psychometrika*, **23**(3), 187-200.

Leaver, J. (2014) The state of Utah's tourism, travel and recreation industry, *Utah Economic and Business Review*, **73**(4), 1-15.

Lobo, R., Goldman, G., Jolly, D., Wallace, D., Schrader, W.L. & Parker, S. (1999) Agritourism benefits agriculture in San Diego County, Reprinted from *California Agriculture*. Available at http://sfp.ucdavis.edu/agritourism/Case_Studies/agritourSD/. (Accessed: 10/09/2017).

Long, L. (2004) *Culinary Tourism*, Lexington, KY: University Press of Kentucky.

Low, S.A., Adalja, A., Beaulieu, E., Key, N., Martinez, S., Melton, A., Perez, A., Ralston, K., Stewart, H., Suttles, S., Vogel, S. & Jablonski, B.B.R. (2015) *Trends in U.S. local and regional food systems*, administrative publication AP-068 prepared for the US Department of Agriculture, Economic Research Service.

Martinez, S., Hand, M., Da Pra, M., Pollack, S., Ralston, K., Smith, T., Vogel, S., Clark, S., Lohr, L., Low, S. & Newman, C. (2010) *Local food systems: Concepts, impacts, and issues*, economic research report ERR-97 prepared for the US Department of Agriculture, Economic Research Service.

Murray, D.W. & O'Neill, M.A. (2012) Craft beer: Penetrating a niche market, *British Food Journal*, **114**(7), 899-909.

National Restaurant Association. (2015) Restaurant industry forecast. Available at https://restaurant.org/research. (Accessed: 19/03/2017).

Nevada Industry Partners. (2018) Visitor Statistics and Profiles. Available at http://nvculture.org/travelnevadabiz/visitor-statistics-and-profiles/. (Accessed: 22/04/2019).

Oom do Valle, P., Correia, A. & Rebelo, E. (2008) Determinants of tourism return behavior, *Tourism and Hospitality Research*, **8**(3), 205-219.

RRC Associates. (2019) Table 2: Estimated U.S. snowsports visits by region, 1978/79 – 2017/18, in 2018-19 Kottke End of Season Survey, report prepared for the National Ski Areas Association, p. 7.

Richards, G. (2011) Creativity and tourism: The state of the art, *Annals of Tourism Research*, **38**(4), 1225-1253.

Rogerson, C.M. (2016) New directions in the South African beer industry: Interpreting the growth of craft beer, *South African Geographers*, **15**, 17.

Rudy, J.T. (2004) Of course, in Guatemala, bananas are better, in L. Long (ed.), *Culinary Tourism*, Lexington, KY: University of Kentucky Press, pp. 131-156.

Sánchez-Cañizares, S. & López-Guzmán, T. (2012) Gastronomy as a tourism resource: Profile of the culinary tourist, *Current Issues in Tourism*, **15**(3), 229-245.

Shenoy, S.S. (2005) Food Tourism and the Culinary Tourist, PhD dissertation, Clemson University, Clemson.

SLC International Airport. (2019). SLC fast facts. Available at www.slcairport.com/about-the-airport/airport-overview/fast-facts/. (Accessed: 22/04/ 2019).

Slocum, S. & Curtis, K. (2017) Farm diversification through farm shop entrepreneurship in the UK, *Journal of Food Distribution Research*, **48**(2), 35-51.

Slocum, S. & Curtis, K.R. (2018) *Food and Agricultural Tourism: Theory and Best Practice*, New York, NY: Routledge.

Smith, S., Costello, C. & Muenchen, R.A. (2010) Influence of push and pull motivations on satisfaction and behavioral intentions within a culinary tourism event, *Journal of Quality Assurance in Hospitality & Tourism*, **11**, 17-35.

Sohn, E. & Yuan, J. (2013) Who are the culinary tourists? An observation at a food and wine festival, *International Journal of Culture, Tourism and Hospitality Research*, **7**(2), 118-131.

StataCorp. (2011) Stata 12 Base Reference Manual, College Station, TX: Stata Press.

Thilmany, D., Sullins, M. & Ansteth, A. (2007) *Of wine and wildlife: Assessing market potential for Colorado agritourism*, Colorado State University Department of Agricultural & Resource Economics Fact Sheet.

US Department of Agriculture Forest Service. (2002) National survey on recreation and the environment.

Utah Economic Council. (2015) *Economic Report to the Governor*. A collaborative endeavour of the David Eccles School of Business and Governor's Office of Management and Budget.

Utah Office of Tourism. (2018) Utah tourism industry metrics. Available at https://travel.utah.gov/research-planning/utah-tourism-industry-metrics. (Accessed: 22/04/2019).

Wilson, J., Thilmany, D. & Sullins, M. (2006) *Agritourism: A potential economic driver in the rural West*, economic development report EDR 06-01 prepared for Colorado State University Cooperative Extension.

Wolf, E. (2014) *Have Fork Will Travel: A Practical Handbook for Food & Drink Tourism Professionals*, Portland, OR: World Food Travel Association.

Yun, D., Hennessey, S.M. & MacDonald, R. (2011) Understanding culinary tourists: Segmentations based on past culinary experiences and attitudes toward food-related behaviour, paper presented at the International Council on Hotel, Restaurant, and Institutional Education conference, 27-30 July, Denver, CO.

Ziesler, P. & Singh, P. (2017) Statistical abstract, report prepared for the US Department of the Interior, National Park Service.

Acknowledgments

This research was supported by the Utah Agricultural Experiment Station, Utah State University, and approved as journal paper number 8877.

11 Governance and Neolocalism: Guiding the Creative Process

Susan L. Slocum

Abstract

The political economy is immensely influential in the way tourism is produced, marketed, and consumed. Governance, and the resulting policy formation and implementation, must be rooted in the locality if it is to support neolocal development. This chapter highlights four case studies from the literature to showcase the role of policy in facilitating or stifling neolocalism within the tourism reality. Examples include national policy (Qatar), regional policy (Kanazawa, Japan), destination marketing and management (Denmark), and the lack of coherent local involvement in policy (Linhares da Beira, Portugal). While some of these policies were not designed with tourism in mind, the chapter supports the role of good governance, specifically the inclusion of local voices and the development of social and human capital, as prerequisites for successful neolocalism.

Introduction

Neolocalism involves the crafting of the tourism product that reflects the culture, history, and value system inherent in a destination. Neolocalism is defined as 'the reaction of individuals and groups to consciously attempt to establish, rebuild, and cultivate local ties, local identities and local economies' (Shortridge, 1996: 10). In other words, neolocalism is the localised pursuit of renewed sense of place through revitalisation and restoration. Flack (1997) recognises this phenomenon as an attempt to reassert the 'distinctively local' (p. 38) in response to a landscape increasingly devoid of the unique, while Brain (2011) highlights the potential of neolocalism in mediating globalisation.

It has been argued that neolocalism can encourage local sourcing, influence environmentally-friendly production techniques and consumption patterns, and encourage support for local causes and charities (Graefe et al., 2018). Evidence suggests that local production can boost the use of local products within the production process, increasing the indirect and induced impacts of tourism. It is not surprising that local products and experiences have become a cornerstone of tourism marketing (Holtkamp et al., 2016) and are increasingly being addressed in tourism policy documentation.

While scant research exists on neolocalism, the focus of limited literature has centred on small businesses, such as breweries, the products they craft, and the growing consumer movement surrounding local consumption (Mathews & Patton, 2016; Murray & Kline, 2015). To date, there has been no specific study that has looked at overall destination development framed by neolocalism. Therefore, this chapter will highlight examples from literature to show how neolocalism is prevalent in destination management (even if not explicitly outlined in such a way) and the strategies needed to successfully fashion destination narratives. In turn, it is hoped that future research will test the conclusions drawn in this chapter and empirically examine the hypothesis that neolocalism is influencing policy implementation and overall destination governance.

Governance

All development occurs within regulated systems of governance. The primary role of governance in tourism is to control growth, sustain communities and environments, and ensure a quality tourism product or experience (Slocum & Curtis, 2017). These objectives are generally addressed through an overarching tourism policy and then implemented at the local level through destination marketing and management organisations or regional tourism boards. Therefore, governance is the leadership structure that coordinates economies, public/private partnerships, and reform objectives designed to pursue collective interests (Slocum & Backman, 2011). Governance can be equated with strategic leadership and includes two primary tasks, government interventions (policy) and managerial tasks (administration), which determine future development paths.

Good governance implies effective policy development that achieves desired outcomes. Agencies have a responsibility to meet the needs of the general public, as opposed to select groups or individuals, and should ensure that power-holders are held to the same standards as the general public. Good governance also implies transparency throughout the policy development process (Bramwell & Lane, 2011). It should be participatory in nature, allow for the equitable distribution of benefits and inclusive decision-making. Dinica (2016) prompts that 'whenever key governance elements are changed, new opportunities and threats emerge; existing unsustainable patterns of development may be reversed or accelerated, and new ones may emerge' (p. 3). Therefore, having local actors intrinsically involved in tourism development is essential to navigate changes to the tourism system and to provide resiliency within a destination.

The reality is that governance agencies are not always effective when developing policy, specifically in the realm of sustainable development. Economic challenges have reduced fiscal support for governing agencies, resulting in more emphasis on industry to lead development agendas (Coles et al., 2012) and find solutions to sustainability issues. As governing agencies transform the role of governance to a role of management and rely on local organisations to fine tune the tourism product, they become more distanced from on-the-ground tourism realities (Pechlaner et al., 2012). Management becomes an avenue to improve destination competitiveness where 'tourism destinations can be regarded as strategic business units' and 'universal solutions may be viewed with scepticism' (Pechlaner et al., 2012: 154-155).

The transformations that are occurring in tourism governance provide new avenues to explore destination competitiveness and the role of local actors in the development process. Hence, as local stakeholders become empowered, they have the ability to influence tourism development in a way that celebrates locality. In other words, they can move from being the object of tourism to the subject of tourism, sometimes known as community-based tourism (CBT) (Lekaota, 2015). CBT puts community at the centre of tourism development, and the way a community responds to tourism development can influence the way policy is implemented, thus, how tourism is managed.

Many scholars recognise the limitations to community involvement in governance, specifically a lack of tourism-based knowledge, limited information on consumer needs and motivations, and an absence of social capital (Slocum & Curtis, 2017; Timothy & Nyaupane, 2009). These limitations can be overcome with the involvement of non-governmental organisations (NGOs) that specialise in small-scale, on-the-ground tourism initiatives. NGOs can provide needed assistance through community involvement of tourism product development and specific expertise through marketing and networking with tour operators (Jänis, 2014). National and global governance bodies have used these agencies to implement sustainable tourism policies and overcome policy barriers (Kline & Slocum, 2015). NGOs have the potential to change the way governments view environments, communities, and tourism as development tools. They also have the expertise to empower local actors and facilitate unique local attributes into the tourism product. Moreover, many NGOs are actively involved in education and training initiatives that encourage innovation and entrepreneurship.

Destination marketing, place-making, and the creative economy

Policy implementation is often accomplished with the help of destination management and marking organisations (DMOs). Slocum and Curtis write, 'Tourism is an industry that must be marketed; as it grows, it must be managed if it is to be sustainable in the long run' (2017: 159). As a highly competitive industry, tourism destinations must differentiate themselves, while simultaneously attracting the right types of businesses to fulfil visitor expectations and advertise to the right mix of potential visitors. Culture has become a crucial aspect of tourism marketing and is one of the fastest growing tourism segments (Richards, 2011). As DMOs attempt to attract culturally minded tourists, expanding the tourism product to include culture-based experiences and ensuring cultural immersion, rather than just cultural sites, is a primary challenge.

Natural and cultural resources help to create and communicate the sense of place, or the special meanings that characterise a destination (Lockie, 2001). Sense of place is established through the physical setting, human activities, the human social and psychological processes, and the experiences associated with the transfer of knowledge about these attributes in a way that communicates place-identity. It is this identity that many tourists seek when choosing a destination, and it is the communication of sense of place that forms the basis of the experiential product and the narrative in tourism. The process of creating, managing, and communicating sense of place is 'place-making' and often falls to DMOs and regional policy-makers to manage. Therefore, DMOs must develop the tourist experience in line with its sense of place and find avenues to authentically represent the essence of a destination if it is to be successful in engaging with cultural tourism.

The creative economy is the production and consumption of imaginative qualities, rather than the use of traditional resources of land, labour, and capital (Howkins, 2002). While the concept of the creative economy was initially used to describe the arts, media, games, and intellectual property, there is a natural connection between creative industries and tourism product development (Richards, 2011). As destinations move away from mass tourism attractions and attempt to differentiate themselves, incorporating creative talents can reinforce place-making and is a valuable resource for neolocalism. Creative businesses are lifestyle enterprises and can involve traditional farms, artisans, crafters, or food and drink producers. The creative economy is a way for creative people to construct and transmit their own evolving identities. Creative industries include (Pivac et al., 2017: 470):

- Advertising
- Architecture
- Crafts and designer furniture
- Fashion clothing
- Film, video and other audio-visual production
- Food and beverage production
- Graphic design
- Educational and leisure software
- Live and recorded music
- Performing arts and entertainments
- Television, radio and internet broadcasting
- Visual arts and antiques
- Writing and publishing

Since tourism is simultaneously produced and consumed as experiences, the mere process of visiting a destination can inherently change a destination's nature. However, the creative economy allows for different types of changes than mass tourism. Rudy (2004) recognises that visitor change includes self-identity and discovery, whereas suppliers find better avenues to enhance their business models and meet the needs of their customers. Creative economies tend to elicit fewer negative personal or community level changes. Creative industries are often small, locally-owned businesses where quality of life is an essential motivator. Moreover, creative entrepreneurs are seeking personal connections with visitors and local bonds with fellow community members (Cavaliere, 2017).

By establishing policies that support product development in line with sense of place, destinations can incorporate neolocalism to offer opportunities for socio-environmental change through a deeper understanding of community values, history, and heritage that connects tourists to community and reconnects people and organisations within communities (Cavaliere, 2017). This process inherently involves recruiting the right combination of businesses that are aligned with place-making narratives (management) and maintaining a quality experience that both motivate visitation and meet customer expectations (marketing). In so doing, governments must develop an industry that facilitates creativity and provides profitable outlets to connect local producers with visitors. In theory, DMOs that engage in neolocal policies will move away from drawing mass tourists and promote a more sustainable tourism type while attempting to foster local ties, local identities and local economies.

Neolocalism as consumption

The influence of globalisation has given rise to a dissatisfaction with mass production and commodified experiences by the traveling consumer. New consumer movements, such as sustainable consumption, LOHAS (Lifestyles of Health and Sustainability), slow, organic, and locavore, show that consumers are increasingly interested in bringing a better quality of life to product producers and minimising the use of natural resources, toxic materials, waste, and pollutants over the product life cycle, so as not to jeopardise the needs of future generations (Seyfang, 2008). There is growing pressure within tourism destinations to meet the needs of tourists interested in unique and authentic visitor experiences.

Moreover, consumers are demanding more sustainable, ethical tourism products. In order to reach this lucrative market, governance organisations are including ethical consumerism in strategic planning and marketing of destinations (Graefe et al., 2018).

The changes occurring throughout the governance process, the rising need for differentiation among destinations, the involvement of NGOs as a support for local participation in the management of tourism, and the need to adapt to consumer pressure for holistic experiences implies that governance structures may have an unintended influence in the rising neolocal movement. It is not surprising that scholars have neglected to address the role of governance in neolocalism as current research into this phenomenon has been limited to consumer movements and small business development. Therefore, a look at the relationship between governance and neolocalism is an important step in establishing best practice and understanding the need to embrace successful governance strategies for neolocal development that supports sustainable tourism.

Best practice in neolocalism

Using four case studies derived from academic literature, this section highlights three examples of how neolocalism is moving into policy implementation at varying levels. The fourth case study highlights how poor policy tools may have unintended side effects that limit neolocalism and create unsustainable development patterns. While not explicitly stated, policy is shown here to be the impetus for certain types of neolocal development and crafted tourism narratives, or policy is presented as being the driving force away from neolocalism. Using examples in national policy frameworks, urban planning, and rural food production, the goal is to call attention to the role policy and governance play in fostering the rising tide of neolocalism.

Art heritage in Qatar – a national policy

According to Weber (2017), the Qatar Tourism Authority, Qatar Museums Authority, and Qatar Foundation for Science and Community Development have established plans to cultivate traditional heritage in the form of revitalising calligraphy, handicrafts, films, books, and the fine arts related to Bedouin culture through the National Vision Strategy 2030 policy document. As a strategy to diversify the economy of Qatar, this policy is specifically designed to enhance the tourism offer of the country and increase its competitiveness as a tourism destination in the Middle East. The policy recognises three specific goals (General Secretariat for Development Planning, 2008: 646):

1 diversifying the economy, which is a natural resources export-based economy;
2 bringing exposure to this little-known area of the world, and;
3 preserving local values

The Qatar Tourism Authority used extensive consultancy methods to utilise international best practices and incorporated a nation-wide consultative process with stakeholders.

Qatar is a country dependent on hydrocarbon sources (oil and gas), and in 2012 only 0.8% of their GDP came from tourism visitation. The primary tourism market was business tourism, including meetings and conventions, and most visitors originated from neighbouring Gulf State countries. Specific development projects to support tourism included a new international airport, built in 2016, and five new museums dedicated to art and history. In preparation for hosting the FIFA 2022 World Cup, investments in tourism infrastructure have included an increase in hotel rooms and attractions, as well as a public metro system.

Traditional Qatari society is based on family and tribal units rather than national identity; therefore, the tourism industry has focused on family-friendly tourism and not 'hedonistic forms of behaviour, such as western-style nightlife' (Weber, 2017: 650). This culture comes from the Bedouin tribes that existed in the country until they were forcibly settled in the 1980s and 90s. As nomadic people, Bedouin artefacts and crafts needed to be small and easily portable and included cloth and rug making, embroidery, camel saddles, hammered metal work, black goat-hair tents, water-bags, and basic jewellery. Traditional foods included dates, sea food, small birds captured by falcons and snares, rice and meat dishes, flat wheat breads, cheese, and fermented milk.

The National Vision Strategy 2030 (2008) includes a number of tools to support neolocalism. Small grants are available to artisans and playwrights. Theatrical performances are produced using local dialects that deal with regional histories and modern concerns. However, film making as a form of creativity has also been supported and has the potential to reach an international audience and support place-making narratives of Qatar. While much of the country was modernised, and old buildings replaced by skyscrapers, a new conservation effort to preserve old architecture is in place and renovations include traditional building styles that promote tangible heritage. The hope is that 'tourists will be able to experience the authentic community living mode' (General Secretariat for Development Planning, 2008: 652) through architecture, art museums, theatres, and local restaurants.

Kanazawa, Japan – an urban policy

Japan has been a hotbed of creativity, in part because of its rapid modernisation after World War II and as a response to the introduction of western values that permeated the internationalisation policies at that time. In Kakiuchi's (2016) study on the creative economy in Kanazawa, Japan, he writes:

> Since then, more attention has been paid to the creation of diversified human capital with talent for innovation...the number of patents, the number of Nobel Prize winners, and the number of scientific research papers published in international journals have been used as measures of creativity (p. 102).

Privatisation and deregulation of the economy in the 1990s has resulted in focused development along the Pacific coast, widening the economic gap across the country. The loss of major industry in outlying cities has weakened local culture, identity, and social cohesion.

Kanazawa was one of the largest cities in Japan during the mid-1800s and was a leading producer of silk and power looms. Because of its remote location, Kanazawa was not selected as an industrial development site after the war, nor was it listed as a protected heritage site under the Law for Protection of Ancient Capitals in 1966. Kanazawa needed to develop its own urban strategy to connect its ancient culture with economic development opportunities in order to increase its financial resources.

The Historical Environmental Preservation Ordinance was introduced in 1968 to control zoning and protect the city's historical environment and was 'the first case of a Japanese regional city enacting regulations for the protection of private houses and the landscape' (Kakiuchi, 2016: 105). Protection measures are in place to preserve old buildings, canals, steep sloping woodlands, temples and shrines, night views, and street scenery. The policy provides for capacity building through the Kanazawa College of Art, workshops, galleries and studios, training courses, and financial assistance for artisans, all resulting in Kanazawa joining the UNESCO creative city network.

As a regional city, attracting tourism has been a difficult task. With the advent of the bullet train, and a link between Kanazawa and Tokyo, tourists are only now discovering the unique cultural and craft heritage of the city. Over 8 million people visited Kanazawa in 2015, and surveys show that cultural assets play a vital role in attracting tourists. Using the concept of design as their primary policy focus, building creative capacity has been at the forefront of their efforts. By developing policies that enhance both human capital and traditional industries, neolocalism has provided opportunities in tourism that would not have existed. Kakiuchi (2016) reminds us that tourism was not the primary focus of the policy, however, it has created opportunities through human capital expansion and small business development to provide visitors with a unique tourist experience and a means to revitalise the economy.

Danish food – destination marketing and management

Gyimóthy (2017) uses the concept of 'terroir' to highlight the role of destination marketing and place-making in Denmark. She recognises the value of destination branding as an avenue to communicate sense of place and to establish a specific place-identity. Highlighting food as the catalyst, she argues that distinct regional culinary specialties only began to appear in the late 1990s in response to the New Nordic Cuisine movement that spread across Scandinavian countries. The enactment of the New Nordic Cuisine Manifesto in 2004, which endorses local, seasonal food foraged freshly from nature or produced sustainably on a small scale, provided support for bottom-up food-based initiatives and collaborative rural development programs. She writes, 'The transformational discourses of New Nordic Food have inspired rural tourism destinations to create a gastronomic sense of place that is distinct from conformist approaches' (p. 1206).

Specifically in the northern periphery of Denmark, two DMOs along the coast (Wadden Sea and the Island of Funen) have capitalised on the remote and isolated landscape imagery to promote 'sweeping, wind-blown beaches, flat marshlands, (and)...the image of a wild, yet amazing rich natural environment' (p. 1207). Slowness as a way of life, as well as organic farming and animal welfare, play a leading role in the marketing messages of the area. Nordic food culture, endorsed by celebrity chefs and televised cooking programs, dominates the marketing materials along with artisanal traditions, rather than labelling or certification, to communicate Nordic identity. Referrals to Viking and medieval heritage have been constructed, although food traditions were only recently embraced in the region (Gyimóthy, 2017).

This narrative has given rise to a unique set of tourism products firmly grounded in concepts of gourmet, natural, cultural-historical, and moral. Through Nordic terroir, market demand for Danish experiences has fuelled rural regeneration and includes new touristic elements that pay close attention to this crafted narrative. The focus is not on the authenticity of existing tourism products, but instead on craftsmanship, lifestyle, and culinary entrepreneurship, and what Gyimóthy (2017) calls 'reversely engineered terroir' (p. 1208). Examples include oyster or beach safaris, gourmet kitchen gardens, medieval markets, visits to local dairies and breweries, and wine, chocolate, and beer culinary routes. However, on a cautionary note, building a destination brand based on consumer preferences and market-demand has led to the development of tourism products that ignore historical recipes and dining traditions. Moreover, the New Nordic Cuisine Manifesto has encouraged a unified Nordic identity that spans Norway, Sweden, and Denmark, diluting individual identities, which could potentially weaken distinctiveness. It is suggested that the use of geographical markers as protected designations should be incorporated into food policy as a means to regulate product development, enhance the destination's authenticity and counter commodification trends inherent in market-driven product development (Gyimóthy, 2017).

Linhares da Beira, Portugal - egalitarianism

Policies designed specifically to increase economic agendas can lead to 'egalitarianism' at the expense of the community, as shown by Chalip and Costa (2012) in rural Portugal. Coles et al. (2012) caution that reduced fiscal support for governing agencies requires more prominence on industry or outside investment by non-local governments to lead tourism development agendas. Top-down planning and trickle-down economics can be counterproductive to neolocal development. Moreover, outside investors rarely understand a community's sense of place and may focus on amenities that clash with local visions, isolate community members from tourism markets, and overemphasise marketing rather than community empowerment (Pechlaner et al., 2012).

Linhares da Beira is located in northern Portugal and has developed a natural and historical tourism-base centred on a walking tour with 26 points of interest (including an open-air museum) and the local castle. Harnessing its natural amenities, the area has also developed robust adventure activities, specifically paragliding. With a long tradition of agricultural production, the district policymakers recognised the potential to increase employment opportunities through rural tourism initiatives. The philosophy was that economic growth would prompt local support for development efforts, create jobs and reduce or reverse rural-to-urban migration (Câmara Municipal de Celorico da Beira 1990–1991). The policy, established in 1990, resulted in approximately €5 million in financial support from the European Union.

Visitor amenities were the priority, including an increase in the number and quality of tourism services, to draw visitors to the region. The plan provided tools to regenerate buildings and homes, which in turn would encourage visitors to walk through town and purchase local handicrafts. However, funding for the project was derived from the Portuguese government, the European Union, and private corporations, resulting in the profits from these investments remaining with the investors outside of the community (Chalip & Costa, 2012).

Without specific training programs in place, local residents were unqualified for many of the new jobs. While classes in local handicrafts were offered as an avenue for local involvement, increasing apathy towards the policy resulted in lacklustre participation. In turn, this lack of support was interpreted as incompetence by the planners. Moreover, the visible wealth disparities between the visitors, the planners, and the entrepreneurs in relation to local residents have further exacerbated the situation. While return migration did occur, these new residents possessed greater entrepreneurial skills and progressive development ideologies, which aligned them with investors and isolated them from long-term community members. Chalip and Costa (2012) write, 'The very fact that entrepreneurs are able to capture benefits confronts egalitarian values, thereby worsening the social climate for tourism development' (p. 42). Mistrust and fatalism have been the end result.

Discussion and future research

This chapter has attempted to highlight the role of policy in the rise of neolocalism, and conversely, the inherent involvement of neolocalism in policy formation. Using four case studies from literature representing national policy, local policy, and the role of destination marketing and management, this chapter has shown that additional research related to neolocalism and policy is warranted. It is the purposeful crafting of identity, sense of place, and the resulting place-making activities that firmly ground three of these case in neolocalism (Flack, 1997). The omission of these elements can lead to fissions between tourism entities and local populations, jeopardising the successful implementation of tourism policy, as shown here in one case study. As noted, neolocalism is not explicitly addressed in any of the cases presented, however, through retrospect, neolocalism has been a foundation of successful policy, and its omission has been a detrimental element.

Slocum and Backman (2011) argue that tourism governance usually occurs through two primary tasks, the writing of policy documents and implementation through a variety of tools, such as regulation, education, investment, and destination management. Qatar provides a solid example of a national tourism policy document that has been used to back neolocal development. The tools used to support this plan include small grants for business, education, and capital investment in museums, airports, and educational facilities. Kanazawa, Japan, on the other hand, has developed an urban policy that protects heritage resources, which has been a major driver for tourism to the area. One policy is purposefully designed for tourism, the other is not, showing how effective policy can serve multiple purposes and accomplish wide ranging development goals (Dinica, 2016). In the case of northern Denmark, policy objectives rely more on the local community and tourism stakeholders and less on government interventions to achieve similar goals (Coles et al., 2012). However, all of the policies are grounded in neolocalism, in that they are purposefully designed to support local identity and rebuild sense of place (Shortridge, 1996).

It is important to recognise the influence of the creative economy in these cases. Whether related to arts, crafts, or film, as in Qatar and Kanazawa, or food in Denmark, using creative talents to implement neolocal policy objectives has been a key driver of success. It is the imaginative quality inherent in creative product development that encourages visitation and supports new forms of tourism visitation (Howkins, 2002). In turn, positive economic provisions for local communities is evidenced, which has the potential to increase the indirect and induced impacts in these destinations (Graefe et al., 2018). Specifically in Denmark, the focus on local and organic gives rise to new outlets for agricultural producers, and in Kanazawa the support of artisans promotes small business development. The oversight of creativity in Portugal's policy has limited the opportunity for residents and created distrust, conflict, and economic leakages. Future research could attempt to measure the contribution of creativity or determine barriers that may be present in accessing consumers interested in the creative economy.

The first three cases also show how a neolocal focus can support place-making and socio-environmental change (Cavaliere, 2017). The emphasis on unique tourism opportunities to encourage growth in the sector has reinvigorated Bedouin heritage and identity in Qatar, after years of suppressive national policies. Moreover, emphasis on film-making provides a new avenue to promote this identity on the international stage. In Denmark, it was the marketing objectives, and the community response to these images, that led to place-making, whereas in Qatar and Kakiuchi, more government interventions, in the form of education and fiscal support, fuelled place-making activities. The emphasis on adventure sports rather than traditional lifestyles has splintered the sense of place and created a dichotomy between visitor perceptions and authenticity in Portugal. Therefore, place-making that is not aligned with a destination can be destructive to sense of place. Destination management necessitates a combination of businesses that assist place-making narratives, and by developing a neolocal branding strategy, the community can use this crafted identity for destination competitiveness (Lekaota, 2015) and small business opportunities. Further investigations should attempt to determine market interest in neolocal tourism products, as well as how interpretive strategies meet customer expectations. Place-making must be communicable and in-line with other destination attributes in the eye of the visitor. They must also align in the view of the community to encourage support and involvement, as shown in Portugal.

Good governance implies that the objectives of policy are actually achieved with minimal unintended impacts (Slocum & Curtis, 2017). Three cases highlight economic diversification through creative industries and place-making. While poverty is not an inherent issue in Qatar

(Weber, 2017), the primary objective of the National Vision Strategy 2030 is to reduce dependence on oil and gas in a world increasingly concerned about greenhouse gas emissions and the changing international political climate (Seyfang, 2008). Time will tell if tourism is a good economic diversification strategy for Qatar. In Kanazawa, the Historical Environmental Preservation Ordinance was designed to protect ancient heritage, specifically tangible architecture and intangible capacity building of artisans. Kakiuchi (2016) ascertains that this has been the catalyst for 8 million visitors, implying a certain level of success in meeting the policy's objectives, both by preserving the city's unique features and developing opportunities for small businesses. However, investments in architecture were insufficient to evoke the same results in Portugal because small business tools were not included in the implementation strategy. The New Nordic Cuisine Manifesto has been instrumental in the rural development of northern Denmark and has spurred creativity through tourism opportunities. However, investigation into who is benefitting and how power is distributed within these destinations could provide valuable insight into whether neolocalism provides a more ethical form of tourism development. One could argue that increased visitation to Portugal signifies a successful tourism industry, yet economic prosperity, the primary goal of their policy, has not been achieved.

One aspect of good governance is the presence of community in the decision-making process. The case studies presented here are silent on the inclusionary processes (if any) conducted on behalf of policy development and implementation. Weber (2017) mentions in passing that Qatar's policy was developed 'following a nation-wide consultative process' (p. 646), but further details are elusive. Kakiuchi (2016) and Gyimóthy (2017) make no mention of inclusion and, rather, refer to community responses after the policies were enacted. Chalip and Costa (2012) specifically highlight the top-down approach, although frequent meetings and attempts at educational programs, however half-hearted, are present. In recognising that these policies were not developed with neolocalism in mind, further research could investigate if community involvement increases support for place-making strategies or if top-down neolocal polices can be equally effective in light of their purposeful intent. If top-down policies have the potential to be successful in a creative economy, what, specifically, is needed to support inclusion. One could argue that the community's adoption of these narratives implies a certain level of support, as governing agencies turn to industry to adequately instigate the implementation of polices (Coles et al., 2012).

Gyimóthy (2017) actively engages in a discussion on commodification and authenticity in relation to 'manipulative rhetoric' (p. 1208) where 'ethics and passion are replacing tradition and history as qualifiers' (p. 1210) in Denmark. Kakiuchi (2016) argues that the use of design can mediate commodification as communities face pressure to add value for visitors. This outlook is supported by Flack (1997) and Brain (2011) who view neolocalism as a significant anti-globalisation strategy. Weber (2017) is quiet on the pressures related to commodification and seems to view Qatar's development as an opportunity for visitors 'to experience the authentic community living model' (p. 652). Chalip and Costa (2012) show that purposeful development is a prerequisite for authentic product development, and that economic growth alone does not constitute an ethical or sustainable industry. Future research should monitor how global pressures influence sense of place and the authenticity of tourism products, as well as community responses that change the nuance of neolocalism. Specific funding mechanisms may inhibit neolocalism.

Lastly, incentives for local involvement is a requirement for neolocalism. Regardless of top-down or bottom-up planning, there must be recognisable benefits for all stakeholders if neolocalism is to produce positive change. While scholars acknowledge the potential for neolocalism to encourage local sourcing and environmentally-friendly production techniques (Graefe et al., 2018), the creative economy should provide an avenue for creative people to negotiate and communicate distinctiveness (Pivac et al., 2017). Identity is a complex concept and transforming it into a consumable asset requires a reassessment of meanings, values, and singularities inherent in a community. How definitions of sense of place are negotiated within heterogeneous communities warrants further investigation.

It is important to note that the roles of NGOs are not integrated into these cases. Their silence cannot be presumed to imply that NGOs were not present in the implementation of these policies, but instead can suggest that their role was not significant or relevant to the points these authors were making. This is a prime avenue for future analysis and could highlight examples of bottom-up planning and community negotiations in the global-local divide and neolocal constructions of identity.

Conclusion

While this chapter has only broached the relationship between neolocalism and policy, it does provide the opening to new conversations around the use and impact of neolocalism in tourism development strategies. Understanding the role of neolocalism, that is the manipulation of identity and the purposeful creation of sense of place narratives, provides a new lens through which to measure and explain sustainable development schemes. Ensuring that locals embrace place-making and find enhanced opportunities through the tourism sector could potentially support more inclusive policy, greater participation at the local level, and new opportunities for visitor and guest relations. It also has the potential to support good governance by building networks between all stakeholders and supporting human capital development in creative tourism products.

References

Brain, T.J. (2011) Examining the Portland music scene through neo-localism, Master's thesis, Portland State University, Portland.

Bramwell, B. & Lane, B. (2011) Critical research on the governance of tourism and sustainability. *Journal of Sustainable Tourism,* **19**(4-5), 411-421.

Câmara Municipal de Celorico da Beira (1990–1991), *Plano de pormenor de recuperação de Linhares da Beira/Detailed Plan for the Revitalization of Linhares da Beira,* Celorico da Beira: Câmara Municipal de Celorico da Beira.

Cavaliere, C.T. (2017) Foodscapes as alternate ways of knowing: Advancing sustainability and climate consciousness through tactile space, in S.L. Slocum & C. Kline (ed.), *Linking Urban and Rural Tourism: Strategies for sustainability,* Wallingford: CABI, pp. 49-64.

Chalip, L. & Costa, C.A. (2012) Clashing worldviews: Sources of disappointment in rural hospitality and tourism development, *Hospitality & Society,* **2**(1), 25-47.

Coles, T., Dinan, C. & Hutchison, F. (2012). May we live in less interesting times? Changing public sector support for tourism in England during the sovereign debt crisis, *Journal of Destination Marketing & Management,* **1**(1-2), 4-7.

Dinica, V. (2017) Tourism concessions in National Parks: neo-liberal governance experiments for a Conservation Economy in New Zealand, *Journal of Sustainable Tourism,* **25**(12), 1811-1829.

Flack, W. (1997) American microbreweries and neolocalism: "Ale-ing" for a sense of place", *Journal of Cultural Geography,* **16**(2), 37-53.

General Secretariat for Development Planning. (2008) *Qatar National Vision 2030.* Available at https://www.gco.gov.qa/en/about-qatar/national-vision2030/. (Accessed: 10/12/2019).

Graefe, D., Mowen, A. & Graefe, A. (2018) Craft beer enthusiasts' support for neolocalism and environmental causes, in S.L. Slocum, C. Kline & C. Cavaliere, (eds.), *Craft Beverages and Tourism, Volume 2 - Environmental, Societal, and Marketing Implications*, Basingstoke, UK: Palgrave MacMillan, pp. 27-48.

Gyimóthy, S. (2017) The reinvention of terroir in Danish food place promotion, *European Planning Studies*, **25**(7), 1200-1216.

Howkins, J. (2002) *The Creative Economy: How people make money from ideas*, Penguin: UK.

Holtkamp, C., Shelton, T., Daly, G., Hiner, C.C. & Hagelman III, R.R. (2016) Assessing neolocalism in microbreweries, *Papers in Applied Geography*, **2**(1), 66-78.

Jänis, J. (2014) Political economy of the Namibian tourism sector: addressing post-apartheid inequality through increasing indigenous ownership, *Review of African Political Economy*, **41**(140), 185-200.

Kakiuchi, E. (2016) Culturally creative cities in Japan: Reality and prospects, *City, Culture and Society*, **7**(2), 101-108.

Kline, C.S. & Slocum, S.L. (2015) Neoliberalism in ecotourism? The new development paradigm of multinational projects in Africa, *Journal of Ecotourism*, **14**(2-3), 99-112.

Lekaota, L. (2015) The importance of rural communities' participation in the management of tourism management: A case study from Lesotho, *Worldwide Hospitality and Tourism Themes*, **7**(5), 453-462.

Lockie, S. (2001) Food, place and identity: consuming Australia's 'Beef Capital', *Journal of Sociology*, **37**(3), 239-255.

Mathews, A.J. & Patton, M.T. (2016) Exploring place marketing by American microbreweries: neolocal expressions of ethnicity and race, *Journal of Cultural Geography*, **33**(3), 275-309.

Murray, A. & Kline, C. (2015) Rural tourism and the craft beer experience: Factors influencing brand loyalty in rural North Carolina, USA, *Journal of Sustainable Tourism*, **23**(8-9), 1198-1216.

Pechlaner, H., Volgger, M. & Herntrei, M. (2012) Destination management organizations as interface between destination governance and corporate governance, *Anatolia*, **23**(2), 151-168.

Pivac, T., Blešić, I. & Kliček, T. (2017) Impact on tourism development–the case study of the creative industries cluster of Vojvodina, *Tourism in Southern and Eastern Europe*, **4**, 469-479.

Richards, G. (2011) Creativity and tourism: The state of the art, *Annals of Tourism Research*, **38**(4), 1225-1253.

Rudy, J.T. (2004) Of course, in Guatemala, bananas are better, in L. Long, (ed.) *Culinary Tourism*, Lexington: University of Kentucky Press, pp. 131-156.

Seyfang, G. (2008) Avoiding Asda? Exploring consumer motivations in local organic food networks, *Local Environment*, **13**(3), 187-201.

Shortridge, J.R. (1996) Keeping tabs on Kansas: Reflections on regionally based field study, *Journal of Cultural Geography*, **16**(1), 5-16.

Slocum, S.L. & Backman, K.F. (2011) Understanding government capacity in tourism development as a poverty alleviation tool: A case study of Tanzanian policy-makers, *Tourism Planning & Development*, **8**(3), 281-296.

Slocum, S. & Curtis, K.R. (2017) *Food and Agricultural Tourism: Theory and best practice*, London: Routledge.

Timothy, D.J. & Nyaupane, G.P. (eds.) (2009) *Cultural Heritage and Tourism in the Developing World: A regional perspective*, London: Routledge.

Weber, A.S. (2017) Creative industries tourism in Qatar: economy diversification through arts and heritage development, *Tourism in Southern and Eastern Europe*, **4**, 645-655.

12 Conclusion

Susan L. Slocum

Introduction

This book, *Neolocalism and Tourism: Understanding a Global Movement*, has attempted to broaden discussion around the current neolocalism movement and its role in tourism development. The term 'neolocal' was coined by Shortridge in 1996, although its vernacular use in tourism studies has remained obscure. Its connections with sense of place, place-attachment, community or regional identity, and storytelling have allowed authors to investigate many of the aspects of neolocalism without using the specific terminology. We can credit Schnell and Reese (2003) with revitalising the concept, which has resulted in its growing use within the tourism literature. Therefore, this comprehensive collection of work expands the theoretical understanding of neolocalism and applies it to a variety of innovative tourism development projects. As stated in our introduction, our goal has been to encourage scholars to 'don a critical lens' and 'explore this book and the case studies within via a vantagepoint of both proposed solutions and evolving contentions' (Cavaliere & Ingram, Chapter 1). While recognising that many of our authors did not conduct research with neolocalism in mind, the case studies presented here have shown the versatility of neolocalism. We have attempted to bring attention to its strengths in understanding tourism development and marketing, as well as many of the social challenges inherent in these activities. Moreover, we hope that our readers recognise neolocalism as a 'tool in the sustainability tool box' (Slocum & Ingram, Chapter 2) rather than as a standalone alternative view of tourism phenomena.

The conclusion is designed to recap the inherent messages throughout the book. By highlighting common themes, the goal is to provide a holistic understanding of neolocalism as presented by our authors. If we recognise neolocalism as 'a way to strengthen insight and communication about, and benefits from, local cultural uniqueness' (Cavaliere et al., Chapter 8) and 'that (which) differentiates the destination and promotes a connected consciousness of the locality' (Leite & Allis, Chapter 7), we inherently understand that neolocalism and its application is as diverse as each community, destination, culture, or place. 'Human bodies are not simply vessels of consciousness' (Carolan, 2009: 1-2) but our embodiment and our senses are critical avenues to multiple transformative ways of knowing, as evidenced throughout the chapters of this book.

Moreover, it is important to recognise the inherent difference between localism and neolocalism. Neolocalism involves the intentionality of using localism, in this case through tourism development and promotion (Shortridge, 1996). Booth-Smith (2017) claims that empathy is absent from localism, as it turns our vision inward without an understanding of the inherent connections each locality has with the rest of the world. Quirk (2017: 9-10) writes,

> We are all tethered to the fabric of locality. This tethering can tie us down through attachments that limit our horizons or our scope for action. For the limit of localism is parochialism. But when properly attached to the fabric of locality we can also develop our capabilities and our confidence to craft new and productive attachments to other places nearby or across the globe (p. 9).

These authors see neolocalism as the freedom to engage with the local as a means to communicate sense of place to a global audience and to embrace the interconnectivity of all localities in shaping human understanding, interactions, and diplomacy. Therefore, neolocalism widens a locality by placing it within global society. In attempting to further our understanding of neolocalism, it is important to craft a definition that embraces these ideologies, to separate neolocalism from localism, and allow for its changing representation as a theory as its use evolves.

Lessons learned

There are a number of themes revealed throughout these chapters. These include: definition versus redefinition of sense of place; heterogeneity of cultural representation; facilitators as actors in the development of neolocalism; community capital inherent in neolocal facilitation; and the consumption of neolocalism. Each of these will be further explained and aspects for future research are provided.

Definition versus redefinition of sense of place

Sense of place is an obvious cornerstone of neolocalism (Holtkamp et al., 2016) and the fundamental construct in the understanding and application of the neolocal framework (Fagence, Chapter 4). Schnell (2013) describes neolocalism as 'a new narrative of place'. Peng et al. (Chapter 9) highlight 'place-identity formation', whereas Derrett (Chapter 6) describes the 'attempt to reinvigorate the local' and Slocum (Chapter 11) offers 'renewed sense of place'. Whether used as a strategy for creation or revitalisation, neolocalism is 'a complex phenomenon with interconnected components grounded in a multi-disciplinary array of theories and concepts' (Ingram, Chapter 3).

Neolocalism requires the intentional construction of inherent value related to locality, which makes it a useful tool for both the construction of identity and the reassessment of locality. It defines authenticity as communities negotiate their identity through change and evolve over time (Richards, 2009). Most importantly, neolocalism allows for place consciousness to be integrated more fully into both the strategic planning and marketing of a destination. For example, Fagence (Chapter 4) uses 'nodes of geographical meaning' to explain different historical narratives surrounding Ned Kelly, the bushranger-cum-outlaw, that have merged into a regional theme where 'sense of place, place-attachment, place-identity, and place-personality may migrate as the telling of the story unfolds'. Aquino and Kloes (Chapter 5) also showcase the evolution of a theme, here the Icelandic sagas, that form the backbone of tourism development. These examples show both the natural evolution of sense of place, as individual towns take stock in their assets related to Ned Kelly and add their individual contributions to the overall tourism product, as well as the purposeful construction of place narrative, where a number of Icelandic communities sought out and developed a place-based brand. These two destinations appear to be rather isolated from other traditional, perhaps mass, forms of tourism requiring a strategic assessment of tourism assets through which sense of place is developed and communicated (Lockie, 2001).

It is the intentionality of neolocalism that makes it malleable and varied across destinations. Unlike other forms of tourism development, local knowledge must form the foundation on which sense of place resides (Lichrou et al., 2008). In turn, sense of place informs the types of tourism developed and the narratives which communicate and engage the consumer. As scholars continue to use the neolocal lens in their research, understanding the fundamental differences between defining or redefining, designing or altering, or developing or revitalising communities and the tourism industry are critical elements for future research. Intentionality implies decision-making, which in turn requires best-practice guidelines for local development stakeholders if neolocalism is to be a sustainable choice. This is an area of research that could provide insight into more ethical tourism products and experiences.

Heterogeneity of cultural representation

Social change is an inevitable process, and neolocalism has the potential to support a community's assessment and acceptance of these changes. In particular, globalisation and mobilities has made cultural heterogeneity the norm rather than the exception. Fagence (Chapter 4) encourages us to 'thinking geographically', which implies that neolocalism must embrace the diversity inherent in any geography. Cavaliere and Ingram (Chapter 1) claim that 'diversity sustains our corporeal, spiritual, and intellectual health and is what drives us to experience the world'. As diverse communities integrate, the need to maintain cultural distinctiveness, yet facilitate what Hall et al. (2012) call 'inclusive growth' is vital to the success of tourism development. Derrett (Chapter 6) shows how neolocalism rediscovers multiculturalism within the confines of national identity, and Slocum and Ingram (Chapter 2) claim that neolocalism is well situated to 'place the power of production into the hands of those most influenced by unsustainable forms of tourism'. Sustainable tourism promotes the maintenance of the world's diversity, both environmental and cultural, and neolocalism can ensure the inclusion of both current cultural representations and cultural manifestations long into the future.

Neolocalism requires connections to local innovation and social entrepreneurship, 'where entrepreneurs play the role of change agents by developing a mission to create and sustain social, and not just private, value' (Hall et al., 2012: 785). Leite and Allis (Chapter 7) provide insights into a community that is attempting to maintain its cultural values in the face of globalisation. Bom Retiro, Brazil, is founded on its Korean heritage, yet has evolved as modern Korean cultural has evolved. The adoption of K-pop music, animes, and emerging food trends show how neolocalism permits the blending of traditional Korean culture with the evolving culture of second-generation Brazilian-Korean youth. The community has recognised 'a constant need to reinforce the ethnic features of the products as authentically Korean through social media feedback loops that monitor consumer responses to change.'

Derrett (Chapter 6) acknowledges that neolocalism 'can be manifest through the experience of groups celebrating their heritage in locations of a diasporic separate from their place of origin'. Her case study highlights how an Italian community helps define national culture through multiculturalism, which supports new relationships and the building of collaboration and cooperation. Slocum and Ingram (Chapter 2) reiterate the message that neolocalism provides opportunities for relationship building. This result reinforces Eades et al.'s (2017) findings that neolocalism can link place-attachments from the neighbourhood and small-town to bind a distinctively national identity. As Cavaliere and Ingram (Chapter 1) point out, neolocalism reinforces 'commodification to both assimilate and maintain cultural distinctiveness'. If neolocalism is a response to global homogeneity, celebrating multiculturalism within diverse communities explains the varied histories that influence the uniqueness of place.

Human migration is a historic phenomenon, and the reality on the ground is that communities are becoming 'melting pots' of ethnic diversity. Future research on how multiculturalism is integrated into neolocal development could provide insight into cultural negotiation, specifically as communities define and market their unique attributes. Schnell (2013) claims that there are 'different ways that local is conceived of by participants in neolocal movements, and the implications that this has on the identities that they derive from it' (p. 65). As scholars critique the neolocal movement, it will be important to consider the inclusion of minorities and those outside the mainstream in order to judge the value of intentionally constructed narratives of self in communities that adopt neolocalism.

Facilitators as actors in the development of neolocalism

Holtkamp et al. (2016: 66) defines neolocalism as 'a conscious effort by businesses to foster a sense of place based on the attributes of their community'. However, the chapters in this book emphasise the multitude of facilitators that enable neolocalism, not just business enterprises. While the role of business cannot be understated, neolocal stakeholders span a wide range of local actors, including local government, community festival organisers, volunteers, and local residents, as well as national policy-makers who support flexibility in local decision-making.

The role of businesses in development is inherently contentious. While Ingram (Chapter 3) acknowledges, 'small businesses have adopted the tenets of CSR, which correlate to the values of the owner(s) and community while gaining a competitive advantage', business is inherently profit oriented and exploitive in nature (Higgins-Desbiolles, 2010). The example of the Silos, in Waco, Texas, suggests business may not be best suited to lead neolocalism, especially when their profit goals encourage a move towards mass tourism and commodification of the very culture that supports their growth. Yet other chapters support the notion that business is vital to the success of neolocalism, such as the entrepreneurial talents in Öland (Peng et al., Chapter 9) and Bom Retiro (Leite & Allis, Chapter 7). In fact, Solvoll et al. (2015: 120), emphasise that, 'the structural change and transition to more experience-based products in tourism demand entrepreneurial behaviour in order to implement needed innovations'. Therefore, it could be said that the singularity of development in Waco is inherently problematic, and that neolocalism requires a more collaborative approach to sense of place development and marketing, with the explicit inclusion of social capital (Parnwell, 2007)

Tourism literature has highlighted the active role non-governmental organisations play in uniting stakeholders around tourism development. Slocum and Klitsounova (2020: 73) write 'NGOs are often the voice of the disenfranchised and have the potential to change the way governments view environments, communities, and tourism as a development tool'. NGOs have been working over the years to fulfil some of the embedded goals of neolocalism. For example, the Lismore's Friendship Festival in Australia began as a volunteer event, but became formalised through the creation of an independent incorporated association to support its growing structural and strategic needs. This organisation works to 'minimise overcrowding... and mitigate the risk of authenticity being diminished as the process of increasing resident pride and cultural identity raises the city's profile' (Derrett, Chapter 6). In Iceland, Aquino and Kloes (Chapter 5) attribute the success of both the Grettir the Strong Project and the Icelandic Seal Center to the formation of non-profit organisations that bring together a number of local stakeholders. Skördefestens Vänner Ideell Förening, the strategic leadership for the Öland's Harvest Festival in Sweden, is also a non-profit, community-based organisation (Peng et al., Chapter 9). Slocum (Chapter 11) argues that these organisations not only educate communities on the advantages of tourism development, but also provide access to funding, networking, and marketing, and can unite governments and communities around common neolocal ideologies.

Neolocalism is fundamentally about society; its sense of place, and its relationship to historical and environmental attributes. Neolocalism embraces personal values, beliefs, and experiences of local (Schnell, 2013). Therefore, future research should remain grounded in the heterogeneity inherent in social science research and in communities, no matter how large or small. While this statement is not revolutionary in nature, neolocalism is inherently based on variety, difference, and uniqueness, requiring researchers of neolocalism to maintain a multifaceted perspective within their research. How communities negotiate diversity of thought is a fundamental aspect of neolocalism, as neolocalism requires a constructed (and hopefully inclusive) narrative of self and others within a common geographical sphere. The role of business is an inherent necessity when developing an economic sector, yet when left to their own devices, businesses alone may not be the best vessel to deliver neolocalism (Ingram, Chapter 3). Perhaps a reassessment of Holtkamp et al.'s (2016: 66) definition of neolocalism as a 'business' endeavour is warranted.

Community capital inherent in neolocal facilitation

All human interactions are inherently grounded in power structures and struggles, and the distribution of power is an ongoing challenge within and between tourism stakeholders. In Chapter 1, Cavaliere and Ingram ask readers to 'further our collective questioning of community versus corporate power and therefore, the need for economic re-appropriation'. Neolocalism is well suited to balance power (Schnell, 2013), as it allows a more organic definition of sense of place and sense of community. Yet, Derrett (Chapter 6) recognises that community festivals, a documented avenue towards neolocalism, may actually mask power relationships, resulting in increased tensions surrounding societal differences. Bianchi (2018: 99) writes, 'cultural political economies of tourism underplay the systemic nature of capitalism as well as the manner in which particular sets of ideas and discourses gain material force'. Therefore, access to capital can inherently shift power structures and empower communities to engage in the direction of their own future. Peng et al. (Chapter 9) shows how neolocalism can foster and facilitate social interactions, enable social replication, and support heritage and modern craftmanship within existing social capital networks.

Access to capital is often a mitigating factor in power relationships. Slocum and Ingram (Chapter 2) write, 'capitalism, or the ability of a few to capitalise on common pool resources at the expense of others, forms the foundation of political sociology and sustainability'. The most obvious example is the Silos in Waco, Texas (Ingram, Chapter 3). The sheer size and investment power of this complex has allowed it to dominate development at the expense of small business and the local community. Cavaliere et al. (Chapter 8) argue that neolocal development should be based on local social and natural capital that augment sense of place and local identity. Yet, the Silos (Chapter 3) is an example where social and natural capital are overshadowed by built and financial capital. Aquino and Kloes (Chapter 5) also show how barriers to financial capital and the loss of other community capitals during the 2008 recession were detrimental to many of the neolocal projects in Iceland.

Yet, neolocalism also strengthens social capital, which appears to be a primary driver in many of the case studies presented in this book. In his foundational article, Shortridge (1996) identifies neolocalism as an opportunity to establish social ties and identities. The positive interactions inherent in Peng et al.'s study of Öland's Harvest Festival (Chapter 9) highlight the value of social capital development. Jones (2005) argues that social capital, in particular trust between stakeholders, is essential to developing other forms of capital. Trust reduces the cost, and increases the reliability, of information and encourages the sharing of ideas throughout the decision-making process. Aquino and Kloes (Chapter 5) support the notion that the perseverance of social capital after the economic recession may have allowed Húnaþing vestra to rebound and reengage in tourism development. Cavaliere et al. (Chapter 8) underscore the presence of social capital in Bangkok as an asset that could support further tourism development along the Saen Saeb Khlong. In order to ensure an inclusive neolocal development model, further investigation into the role of social capital on successful neolocal development is needed. The case of the Silos (Ingram, Chapter 3) shows that neolocal development without social capital may inherently result in mass tourism development.

The foundation of neolocalism is that it should support positive interactions and require community involvement. Derrett (Chapter 6) writes, 'personal and professional local inputs (experience, wisdom, skills, and special interests) can deliver powerful 'bottom up' stories that can inform and empower participants'. Moreover, community resiliency is also a common theme throughout this book. While several authors have emphasised the need for local decision-making as a prerequisite for successful neolocal development, Slocum (Chapter 11) writes, 'investigation into who is benefitting and how power is distributed within these destinations could provide valuable insight into whether neolocalism provides a more ethical form of tourism development'. The role of neolocalism in policy development, as well as the implementation of policy at the local level should be addressed in future research.

Consumption of the neolocal

Curtis et al. (Chapter 10) acknowledge that diversity is vital to attracting visitors. However, the inherent nature of neolocalism lies in the ability to consume the local, which is a 'negotiated phenomena' (Derrett, Chapter 6) and requires feedback between communities and the tourists. Slocum and Ingram (Chapter 2) argue that 'it is the creative process inherent in the production of cultural tourism products that enhances and supports traditional production and appeals to tourism consumers'. Peng at al. (Chapter 9) assert that it is the spatial overlaps between tourists, visitors, and locals, which generates vitality and diversity and 'incidental visibility' merging both global and local trends for consumption.

Consumers are drawn to neolocal experiences as a way to form deeper connections with destinations (Rudy, 2004). Yet, the very nature of tourism often results in the commodification of tourism experiences. Fagence (Chapter 4) writes, 'whereas the baseline characteristics of neolocalism may be inward-looking, locality-supporting, and locality-protecting, those of tourism are almost always outward-looking, locality-maximising, and attention-seeking'. Kakiuchi (2016) believes that the intentional construction inherent in neolocalism can counter commodification, or at least place the power of product design into the hands of the community, and Chalip and Costa (2012) argue that it is the purposeful development of place that maintains authentic product development. Therefore, future research could focus on how authenticity and commodification are negotiated within neolocalism, both among community members, as well as between the local production and consumption of tourism products. Moreover, successful neolocal development could result in overconsumption, implying that successful neolocalism may alter or destroy authenticity.

Curtis et al. (Chapter 10) write 'understanding tourists' characteristics, motives, and behaviours is key to attracting them', therefore, further research is needed to better comprehend the connections between sense of place marketing and neolocal consumption. They found that all four clusters in their study harboured elements of neolocalism, implying that neolocalism may have a broad reach to a variety of tourism typologies. Graefe et al. (2018) found that supporters of neolocalism were more likely to seek out personal connections, both close friendships and casual exchanges, and to form social groups through blogs and websites. This result implies that neolocal consumers may carry the discovered sense of community to other aspects of their lives, including the building of new (virtual) communities that support and encourage further consumption of neolocalism. Schnell (2011), explains, 'the promotion of these connections is...also driven by travellers' wish to see and experience the world as one ruled not by impersonal networks of capital, but rather by networks of local, personal connections' (p. 302).

If neolocalism is to support authentic tourism experiences, garner inclusive development, and support economic development goals, research must provide insight on the co-creative processes that support neolocalism. It is unlikely that the consumer of tourism knows the difference between localism and neolocalism, therefore investigations into the reception of neolocal messages and narratives can support the viability of commercialising localism. Moreover, the impact of tourism consumption on neolocal production is a worthy research agenda and can support best-practice for communities engaging in neolocalism. Cavaliere and Ingram (Chapter 1) warn that 'this focus on consumption is precisely what makes neolocalism vulnerable to the abuses of over consumption by putting those with the means and power in the position to dominate local interests, shutting out those who do not have the means and power to participate'.

More research

The editors of this book are hopeful that scholars will continue to view tourism development through the lens of neolocalism. These case studies have shown, like many sustainable tourism tools, that it is not *what* is done to develop tourism, but rather *how* it is done that influences the success of neolocalism. Simplistic engagement with sense of place, as shown by the Silos, is not enough to warrant a truly sustainable tourism product. Neolocalism inherently lies with the collaborations and social capital that supports multifaceted development models grounded both in community identity, sense of place, and multiculturalism (Parnwell, 2007).

There are many aspects of neolocalism that are not included in this book, which provides ample opportunity for future scholars. Holtkamp et al. (2016) includes three primary indicators: (1) use of local names and images in labelling and marketing; (2) environmental sustainability; and (3) social and community engagement. While the chapters in this book have advanced the scholarly recognition of neolocalism in tourism, these three points are remarkedly absent from this publication and require further investigation.

Neolocalism is grounded in marketing, yet there is little research on the use of sense of place in the marketing of neolocalism in this book. Curtis et al. (Chapter 10) argue that 'destination marketing messages encourage tourists to spend their budgets in a way that reduces economic leakages' and that research should develop targeted marketing plans of neolocal place-based products that best suit tourists. While craft beer research has addressed the role of neolocalism in marketing (see Gyimóthy 2017; Graefe et al., 2018; Murray & Kline, 2015), the function of constructed narratives in attracting visitors should be applied to other tenets of the tourism product. How place names are used in festivals, trails, clusters, or other non-food related travel promotion is important in understanding the viability of neolocalism in attracting customers.

Positioning neolocalism within the realm of the triple bottom line of sustainability is also not well developed in these chapters. While community engagement and social sustainability is well documented, how neolocalism mitigates environmental and economic challenges on the ground needs further investigation. 'The core of neolocalism and its focus on such 'mainstream' issues as quality of life, lifestyle, environment, and 'least disturbance' from outside forces' (Fagence, Chapter 4) could be better assessed. Cavaliere and Ingram (Chapter 1) see neolocalism as a proponent to curb climate change, and Cavaliere et al. (Chapter 8) argue that it can 'serve as conduit for both locals and tourists alike to embrace...biocultural conservation'. Therefore, how the intentionality of neolocalism supports these processes is warranted.

Slocum (Chapter 11) notes that 'localised pursuit' is a key aspect to the definition of neolocalism, and therefore requires policies that allow decision-making at a local level. Yet, how those decisions are negotiated, the wat systems of interactions are implemented, and which narratives are selected for communication are not highlighted in this book. Chalip and Costa (2012) recognise 'clashing worldviews' as a struggle inherent in tourism development. Using the principles of neolocalism may help communities address inherent societal differences related to age, background, ethnicity, and gender and can support a more collaborated approach to definitions, and in turn applications, of sense of place. Moreover, how policies are implemented at the local level can inform neolocal development. Therefore, further investigation, specifically grounded in neolocalism, can inform how local social networks include or omit narratives of diversity. The role of entrepreneurs is a singular aspect of neolocalism that may contradict inclusive decision-making.

Lastly, this book is overwhelmingly qualitative in nature, as is common when exploring new phenomena (Richards & Morse, 2007), culture, marketing (Moisander & Valtonen, 2006), or community-based assessments in their natural environments (Mitchell & Jolley, 2012). As foundational knowledge of neolocalism is expanded, there may be opportunities to quantify some neolocal elements. Understanding the impacts of neolocalism on issues such as economic growth, social capital enhancement, policy implementation, and quality of life indicators can support best practice in neolocal development and marketing.

Conclusion

The societal shift towards neolocalism, and the theoretical underpinning inherent in the socially constructed and communicated sense of place and sense of community, makes neolocalism a multifaceted theoretical construct that appears to be well suited for sustainable tourism research and product development. As we submit this book to our publisher, the world is on lockdown, economies are collapsing, and tourism is hard hit by the unfolding COVID-19 global pandemic. As we emerge from this tragedy into a 'new normal', we see neolocalism as a relevant and optimistic vein of study that could potentially support new, more ethical avenues for tourism development at the local level. We hope that scholars will continue this vein of examination as a means to better understand the necessary elements of inclusive development and as a way to bring best-practice guidelines to practitioners. Tourism production and consumption is a complex occurrence, and the application of neolocalism to include the multifaceted nature of tourism can develop and expand its application. The hope is that as global systems become localised, the locus of control can remain with those most impacted by unsustainable practices.

References

Booth-Smith, L. (2017) What is neo-localism?, in L. Booth-Smith (ed.), *Neo-localism Rediscovering the Nation*, London: Localis, pp. 6-7.

Bianchi, R. (2018) The political economy of tourism development: A critical review, *Annals of Tourism Research*, **70,** 88-102.

Carolan, M.S. (2009) 'I do therefore there is': enlivening socio-environmental theory, *Environmental Politics*, **18**(1), 1-17.

Chalip, L. & Costa, C.A. (2012) Clashing worldviews: Sources of disappointment in rural hospitality and tourism development, *Hospitality & Society*, **2**(1), 25-47.

Eades, D., Arbogast, D. & Kozlowski, J. (2017) Life on the "Beer Frontier": A case study of craft beer and tourism in West Virginia, in C. Kline, S.L. Slocum & C.T. Cavaliere (eds.), *Craft Beverages and Tourism, Volume 1*, pp. 57-74, Cham: Palgrave Macmillan.

Graefe, D., Mowen, A. & Graefe, A. (2018) Craft beer enthusiasts' support for neolocalism and environmental causes, in S.L. Slocum, C. Kline & C.T. Cavaliere (eds.), *Craft Beverages and Tourism, Volume 2*, Cham: Palgrave Macmillan, pp. 27-47.

Gyimóthy, S. (2017) The reinvention of terroir in Danish food place promotion, *European Planning Studies*, **25**(7), 1200-1216.

Hall, J., Matos, S., Sheehan, L. & Silvestre, B. (2012) Entrepreneurship and innovation at the base of the pyramid: A Recipe for inclusive growth or social exclusion?, *Journal of Management Studies*, **49**(4), 785-812.

Higgins-Desbiolles, F. (2010) The elusiveness of sustainability in tourism: The culture-ideology of consumerism and its implications, *Tourism and Hospitality Research*, **10**(2), 116-129.

Holtkamp, C., Shelton, T., Daly, G., Hiner, C.C. & Hagelman III, R.R. (2016) Assessing neolocalism in microbreweries, *Papers in Applied Geography*, **2**(1), 66-78.

Jones, S. (2005) Community-based ecotourism: The significance of social capital, *Annals of Tourism Research*, **32**(2), 303–324.

Kakiuchi, E. (2016) Culturally creative cities in Japan: Reality and prospects, *City, Culture and Society*, **7**(2), 101-108.

Lichrou, M., O'Malley, L. & Patterson, M. (2008) Place-product or place narrative(s)? Perspectives in the marketing of tourism destinations, *Journal of Strategic Marketing*, **16**(1), 27-39.

Lockie, S. (2001) Food, place and identity: consuming Australia's 'Beef Capital', *Journal of Sociology*, **37**(3), 239-255.

Mitchell, M.L. & Jolley, J.M. (2012) *Research Design Explained* (8th edn.), Belmont, CA: Wadsworth.

Moisander, J. & Valtonen, A. (2006) *Qualitative Marketing Research: A Cultural Approach*, London: Sage.

Murray, A. & Kline, C. (2015) Rural tourism and the craft beer experience: Factors influencing brand loyalty in rural North Carolina, USA, *Journal of Sustainable Tourism*, **23**(8-9), 1198-1216.

Parnwell, M.J. (2007) Neolocalism and renascent social capital in northeast Thailand, *Environment and Planning D: Society & Space*, **25**(6), 990–1014.

Quirk, B. (2017) Why place matters, in L. Booth-Smith (ed.), *Neo-localism Rediscovering the Nation*, London: Localis, pp. 7-10.

Richards, G. (2009) Tourism development trajectories — from culture to creative?, paper presented at the conference The Asia - Pacific Creativity Forum on Culture and Tourism, 3-5 June, Jeju Island, Republic of Korea.

Richards, L. & Morse, J.M. (2007) *User's Guide to Qualitative Methods*, 2nd edn, London: Sage.

Rudy, J.T. (2004) Of course, in Guatemala, bananas are better, in L.M. Long (ed.), *Culinary Tourism*, Lexington, KY: University of Kentucky Press, pp. 131-156.

Schnell, S.M. (2011) The local traveler: Farming, food, and place in state and provincial tourism guides, 1993–2008, *Journal of Cultural Geography*, **28**(2), 281-309.

Schnell, S.M. (2013) Deliberate identities: becoming local in America in a global age, *Journal of Cultural Geography*, **30**(1), 55-89.

Schnell, S.M. & Reese, J.F. (2003) Microbreweries as tools of local identity, *Journal of Cultural Geography*, **21**(1), 45-69.

Shortridge, J.R. (1996) Keeping tabs on Kansas: reflections on regionally based field study, *Journal of Cultural Geography*, **16**(1), 5-16.

Slocum, S.L. & Klitsounova, V. (2020) Rethinking tourism in Belarus: The opening of a rural economy, in P. Wiltshier, & A. Clarke (eds.), *Community-Based Tourism in the Developing World*, London: Routledge pp. 67-80.

Solvoll, S., Alsos, G.A. & Bulanova, O. (2015) Tourism entrepreneurship – Review and future directions, *Scandinavian Journal of Hospitality and Tourism* **15**(1), 120–137.

Index

www.ingramcontent.com/pod-product-compliance
Ingram Content Group UK Ltd.
Pitfield, Milton Keynes, MK11 3LW, UK
UKHW021828100926
8556IPUK00003B/224